**National
Fire Protection
Association**

ENGINE COMPANY FIREGROUND OPERATIONS

HAROLD RICHMAN
Fire Chief (Retired)
Past President
International Society of Fire Service Instructors

JONES AND BARTLETT PUBLISHERS
Sudbury, Massachusetts
BOSTON   TORONTO   LONDON   SINGAPORE

Jones and Bartlett Publishers

World Headquarters
40 Tall Pine Drive
Sudbury, MA 01776
978-443-5000
info@jbpub.com
www.jbpub.com

Jones and Bartlett Publishers Canada
6339 Ormindale Way
Mississauga, Ontario L5V 1J2
Canada

Jones and Bartlett Publishers International
Barb House, Barb Mews
London W6 7PA
United Kingdom

National Fire Protection Association

1 Batterymarch Park
Quincy, MA 02169-7471
www.NFPA.org

Jones and Bartlett's books and products are available through most bookstores and online booksellers. To contact Jones and Bartlett Publishers directly, call 800-832-0034, fax 978-443-8000, or visit our website www.jbpub.com.

Substantial discounts on bulk quantities of Jones and Bartlett's publications are available to corporations, professional associations, and other qualified organizations. For details and specific discount information, contact the special sales department at Jones and Bartlett via the above contact information or send an email to specialsales@jbpub.com.

Production Credits

Chief Executive Officer: Clayton E. Jones
Chief Operating Officer: Donald W. Jones, Jr.
President, Higher Education and Professional Publishing: Robert W. Holland, Jr.
V.P., Sales and Marketing: William J. Kane
V.P., Production and Design: Anne Spencer
V.P., Manufacturing and Inventory Control: Therese Connell
Publisher, Public Safety Group: Kimberly Brophy
Senior Acquisition Editor: William Larkin

Editor: Jennifer S. Kling
Production Supervisor: Jenny Corriveau
Photo Researcher: Lee Michelson
Director of Marketing: Alisha Weisman
Cover Image: © Glen E. Ellman
Composition: Omegatype, Inc.
Text Printing and Binding: Courier
Cover Printing: Courier

The procedures and protocols in this book are based on the most current recommendations of responsible sources. The NFPA and the publisher, however, make no guarantee as to, and assume no responsibility for, the correctness, sufficiency, or completeness of such information or recommendations. Other or additional safety measures may be required under particular circumstances.

Additional illustration and photographic credits appear on page 192, which constitutes a continuation of the copyright page.

Library of Congress Cataloging-in-Publication Data

Persson, Steve.
  Engine company fireground operations / Steve Persson, Harold Richman. — 3rd ed.
      p. cm.
  National Fire Protection Association.
  Rev. ed. of: Engine company fireground operations / Harold Richman. 1986.
  Includes index.
  ISBN-13: 978-0-7637-4495-3
  ISBN-10: 0-7637-4495-6
  1. Fire extinction—Handbooks, manuals, etc. 2. Engine companies—Handbooks, manuals, etc. I. Richman, Harold. Engine company fireground operations. II. Title.
  TH9310.5.R5 2007
  628.9'25—dc22
  6048   2006037060

Printed in the United States of America
11 10 09 08 07   10 9 8 7 6 5 4 3 2 1

Harold Richman
November 9, 1926–October 16, 1999

A TRIBUTE TO THE AUTHOR AND HIS LEGACY

John R. Leahy, Jr.

Hal started his legacy under unusual circumstances as a volunteer with the "War Time Preparedness Civil Defense Program" in our nation's capital. The number one priority was engine and truck company operations, training under suspected air raid activities similar to the British Fire Brigade.

When the opportunity arose, he joined the United States Marine Corp until the end of the war. He became a member of the newly formed, largest all-civilian fire department of the Federal Civil Service—the 14th District Pearl Harbor Fire Department with 21 fire companies, three crash crews, and two marine fire boats with Hal serving as a Captain, Training Officer, and District Chief.

Eventually Hal returned to the mainland as a fire fighter in Memphis, Tennessee. He was promoted to training officer and then accepted the challenge to expand recruit training. The need was to fill the personnel numbers with people exposed to the latest technology and skills, so he reviewed programs across the country to put in place a quality program. He helped manage these changes as a Shift Commander and implemented a rescue company operation. Subsequent to that, Silver Springs, Maryland, convinced Hal to take over their training and operations program for standardization of heavy commercial and residential growth.

Hal's life was built upon the fundamentals of training; and he was constantly preaching, to anyone who would listen, from the basics to the university level.

He eventually accepted the first career fire chief position of Fairfax County, Virginia, where his concern was for excellence in performance, based on training. Here he was able to implement his dreams one step at a time. From there he went on to become Chief of Prince George County, Maryland, qualifying over 500 officers in the service at the University of Maryland Fire Officer Staff and Command School. Hal also served through the chairs of the International Society of Fire Service Instructors, the National Fire Protection Association, and others; and authored two books that successfully reduced injuries, loss of life, property, and environment. He was a true example of talking the talk, and walking the walk.

Hal was one of the great ones who influenced all those he came in contact with and who inspired them to feel they should influence others as well. I'm proud to say I really knew him personally, and I hope I have done him justice.

Brief Contents

Contents

The National Fire Protection Association and Jones and Bartlett Publishers are pleased to bring you *Engine Company Fireground Operations, Third Edition*. This new edition combines current content with dynamic features for both the instructor and the student.

Engine company personnel are a key part of firefighting operations at the fireground. This book emphasizes the point that fire fighters performing engine company tasks must be properly trained, possess the proper equipment, and be adequately staffed.

Engine Company Fireground Operations, Third Edition covers the basic objectives of engine company work including

the proper supply and use of water to fight fires, and emphasizes that the engine companies should be focused on three major tactical priorities on the fireground: life safety, extinguishment, and property conservation. Other areas of importance covered in this book include protecting exposures, confining the fire, and carrying out overhaul operations.

The *Third Edition* will help reinforce and expand on the essential information and make information retrieval a snap by utilizing the following features:

Learning Objectives
Identify what students should take away from each chapter.

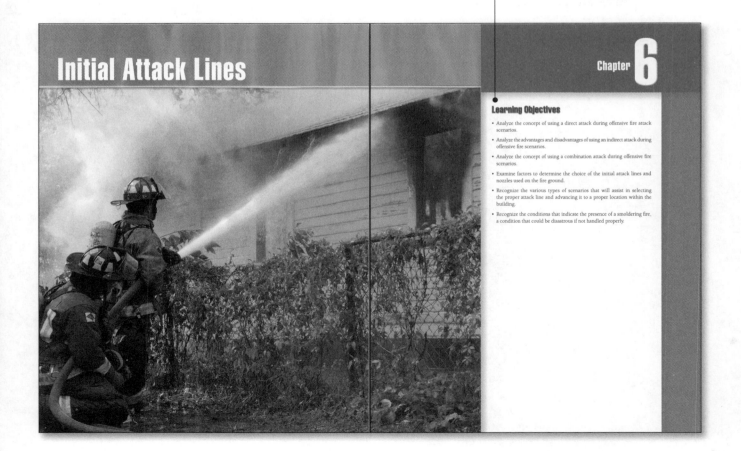

Initial Attack Lines

Chapter **6**

Learning Objectives

- Analyze the concept of using a direct attack during offensive fire attack scenarios.
- Analyze the advantages and disadvantages of using an indirect attack during offensive fire scenarios.
- Analyze the concept of using a combination attack during offensive fire scenarios.
- Examine factors to determine the choice of the initial attack lines and nozzles used on the fire ground.
- Recognize the various types of scenarios that will assist in selecting the proper attack line and advancing it to a proper location within the building.
- Recognize the conditions that indicate the presence of a smoldering fire, a condition that could be disastrous if not handled properly.

Figure 6-2 In an indirect attack, a fog stream is directed at the ceiling of the intensely heated area in order to create steam.

Key terms
Key terms and definitions are emphasized throughout the text.

toward the floor where they are operating. When the area has been cooled down and ventilation has taken place, the operation should be changed from an indirect attack to a direct attack.

Combination Attack

The **combination attack** uses both the direct and indirect methods one after the other. If the room is extremely hot and nearing flashover conditions, the indirect attack is used to bring down the temperature. After this has been accomplished, the direct attack is used to extinguish the main body of fire. A limited amount of water should be used during the indirect attack to limit the amount of steam produced, which may force fire fighters out of the area. Before fire fighters enter the building to conduct firefighting tasks, a risk versus benefit analysis should be conducted.

Sometimes responding companies will find that a fire is smoldering rather than free burning. The exact position of the fire will not be evident, as no flames will be showing. The sequence of operations to be used against a smoldering fire is different from that used against a free-burning fire. The differences are covered in the last section of this chapter.

Keep in mind that an aggressive interior attack should not be made in buildings that are in various stages of demolition, have been abandoned for long periods, have been burning for a predetermined period of time, contain construction features detrimental to safe interior operations, have had previous fires, or are under construction. The IC should conduct a risk versus benefits analysis before the decision is made to allow fire fighters into a building to conduct firefighting activities, including a primary search.

If any interior operations are to be made in these buildings, they should be carried out only after the fire has been knocked down from the outside and a careful check has been made on the condition and relative safety of the structure on the inside.

The normal aggressive interior attack should be made on those buildings that are in use and especially those that are

occupied at the time of the fire. Even in this case, however, if a large intense fire is encountered, it may be necessary to knock down or control the fire from the outside using solid or straight streams before an interior attack can be made.

Remember that occupants will benefit most by the extinguishment of the fire. If the fire is of such intensity that an interior attack cannot control or extinguish the fire with a sufficient amount of attack lines that are appropriately sized, then a defensive attack should be conducted from the outside. If an interior attack is to be made on a free-burning fire, several other decisions will need to be made.

Choosing Attack Lines

One size of hose or one type of stream is certainly not the answer to every fire situation that confronts an engine company on arrival at the fire ground. The choice of initial attack lines and nozzles depends on the purpose of the attack, whether it be a holding action, exposure protection, a defensive operation, or an offensive operation with an interior attack on the main body of the fire.

Factors that affect this choice include the size and location of the fire, how the attack lines are to be used against the fire, available equipment, and the personnel available for fire attack. Based on these factors, the following decisions must be made:
- What size hose diameter is appropriate?
- What type of nozzle should be used?
- How many hose lines are needed?
- Where will hose lines be positioned?
- In what type of operation, offensive or defensive, will the hose lines be used?

Sizes of Attack Lines

Choosing the size or sizes of attack lines depends on the extent and location of the fire and how it will be fought. Most fire

Key Points
Key points are highlighted throughout the chapter to help emphasize important topics.

Wrap-Up
Chief concepts, key terms, and summaries are provided at the conclusion of each chapter.

Fire Fighter in Action
Provide end-of-chapter questions to help students prepare for exams.

Wrap-Up

Chief Concepts

- At a working structure fire, the placement and operation of initial attack lines protect occupants and fire fighters as well as providing the first water on the fire for extinguishment.
- The IC should conduct a risk versus benefit analysis before the decision is made to allow fire fighters to enter a building to conduct an aggressive interior attack.
- The effectiveness of an initial attack depends on several decisions that must be made in rapid succession.
- This decision making begins when the first company arrives on the fire ground and the company officer assumes command.
- A decision needs to be made as to the mode of operation used, offensive or defensive.
- A decision also has to be made regarding the size of hose line to be deployed during the initial attack, how many hose lines are needed, and where they should be positioned.
- The type of nozzle must also be considered, as it will determine the flow and shape of the stream.
- Spray nozzles operate with a stream pattern from straight stream to wide-angle fog and smooth-bore nozzles operate with a solid-stream pattern.
- During an offensive fire attack, an engine company must also consider whether to use a direct or indirect attack.
- A direct attack is one in which a solid or straight hose stream is used to deliver water directly onto the base of the fire.
- An indirect attack is one in which a solid, straight, or narrow fog stream is used to direct water at the ceiling to cool superheated gases in the upper levels of the room with the objective of

preventing flashover, converting the water to steam, absorbing heat, and extinguishing the fire.
- Another important decision concerns the task of ventilating the building.
- Ventilation must be managed with suppression efforts, and a coordinated fire attack, supervised by the IC, will ensure that both tasks take place simultaneously.
- If the building is free burning, the building should be ventilated when the initial attack begins or as soon as possible thereafter.
- If it is a smoldering fire, however, the building must be ventilated before initial attack. In particular, a smoldering fire should be ventilated before any attempt is made to enter the fire building to prevent a backdraft explosion.
- After ventilated, a smoldering fire will burn freely and may be attacked using fire department standard operating guidelines.

Key Terms

Combination attack: A type of attack employing both the direct and indirect attack methods.

Direct attack: Firefighting operations involving the application of extinguishing agents directly onto the burning fuel.

Indirect attack: Firefighting operations involving the application of extinguishing agents to reduce the buildup of heat released from a fire without applying the agent directly onto the burning fuel.

Fire Fighter in Action

1. When selecting the initial attack line:
   a. Select one of the preconnected hose lines.
   b. Select the largest preconnected hose line.
   c. Select the smallest preconnected hose line.
   d. Select either a preconnected hose line or hose line using hose from the hose bed.

2. An attack where water is directed in such a way as to create steam to absorb a large quantity of heat is a(n)
   a. Direct attack
   b. Indirect attack
   c. Absorption method
   d. Heat-reduction method

3. It is recommended that the 1½-inch hose line be
   a. Used only for fires contained to a small room
   b. Used only for all fires in residential buildings
   c. Used when the required rate of flow is 100 gpm or less
   d. Eliminated for structure firefighting

4. A 1¾-inch hose line can discharge _____ gpm.
   a. 60 to 125
   b. 120 to 200
   c. 125 to 150
   d. A maximum of 350

5. When a single 1¾-inch hose line is ineffective in extinguishing a larger fire
   a. Interior operations should be abandoned in favor of a defensive attack.
   b. Additional 1¾-inch hose lines should be deployed until the total quantity of water is sufficient to control the fire.
   c. Additional larger hose lines should be used.
   d. All of the above are options when faced with a situation where a single 1¾-inch line is insufficient.

6. Smooth-bore nozzles attached to the same size hose line
   a. Have greater reach than spray nozzles
   b. Will flow more water than spray nozzles
   c. Have greater reach and will flow more water than spray nozzles
   d. Are approximately equal to spray nozzles in reach and flow capacity

7. _____ stream(s) is/are the safest and most effective when conducting interior firefighting operations in an occupied area.
   a. Solid or straight
   b. Sixty-degree angle fog
   c. Wide-angle fog
   d. Intermittent straight and fog

8. A _____ stream is most effective in pushing the fire.
   a. Solid or straight
   b. Sixty-degree angle fog
   c. Wide-angle fog
   d. Intermittent straight and fog

9. When confronted with a well-involved fire in a part of a structure below an occupied area (e.g., basement storage area fire below apartments), it is usually best to
   a. Immediately rescue all visible occupants.
   b. Attack the fire using a 1¾-inch line with automatic nozzle, and lay a 1¾-inch line to the floor above the fire.
   c. Attack the fire using a 1¾-inch line with solid stream nozzle, and lay a 1¾-inch line to the floor above the fire.
   d. Attack the fire using a 2½-inch line with solid stream nozzle, and lay a 1¾-inch line to the floor above the fire.

10. Of the methods listed, advancing hose via _____ will require the most hose to advance a line to an upper floor of a structure.
    a. The stairs
    b. A ladder
    c. A rope
    d. A pike pole

11. A _____ fire must be ventilated before an offensive attack is begun.
    a. Free-burning
    b. Postflashover
    c. Smoldering
    d. All fires must be ventilated before an interior attack is made.

12. Backdraft is defined as
    a. The rapid ignition of all contents and fire gases in an area
    b. Fire gases suddenly igniting overhead
    c. A sudden explosive ignition of fire gases
    d. An oxygen-starved fire

Instructor Resources

A complete teaching and learning system developed by educators with an intimate knowledge of the obstacles you face each day supports *Engine Company Fireground Operations, Third Edition*. The resources provide practical, hands-on, time-saving tools like PowerPoint presentations, customizable lecture outlines, test banks and image/table banks to better support you and your students.

Instructor's ToolKit CD-ROM

ISBN-13: 978-0-7637-5234-7

ISBN-10: 0-7637-5234-7

Preparing for class is easy with the resources on this CD-ROM, including:

- **Adaptable PowerPoint Presentations**—Provides you with a powerful way to create presentations that are educational and engaging to your students. These slides can be modified and edited to meet your needs.
- **Lecture Outlines**—Provides you with complete, ready-to-use lecture outlines that include all of the topics covered in the text. Offered in word documents, the lecture outlines can be modified and customized to fit your course.
- **Electronic Test Bank**—Contains multiple-choice and scenario-based questions, and allows you to originate tailor-made classroom tests and quizzes quickly and easily by selecting, editing, organizing, and printing a test along with an answer key, including page references to the text.

   The resources found on the *Instructor's ToolKit CD-ROM* have been formatted so that you can seamlessly integrate them onto the most popular course administration tools.

Acknowledgements

Reviewers

Thomas W. Aurnhammer
Los Pinos Fire Protection District
Ignacio, Colorado

Michael Brown
Wilson Technical Community College
Wilson, North Carolina

John A. Cannon
Portland Fire Department Academy
Portland, Maine

Ernie Close
City of Ann Arbor Fire Department
Pinckney, Michigan

Scot Cullers
Loudon County Department of Fire-Rescue
Leesburg, Virginia

Ken Farmer
Former Fire Chief
Fuquay-Varina Fire Department
Fuquay-Varina, North Carolina

Todd Gilgren
Arvada Fire Protection District
Arvada, Colorado

Craig W. Huntsinger, Senior
Department of Fire and Rescue Services (Retired)
Reading, Pennsylvania

Michael Nelson
Elk Grove Township Fire Department Training Academy
Arlington Heights, Illinois

Paul C. Pullar
Walden Fire Department
Walden, New York

John Schuldt
Illinois Fire Chiefs Association
Carpentersville, Illinois

Marty Walsh
San Diego Miramar College
San Diego, California

Ralph Woodbury
Monroe Fire Department
Monroe, Michigan

Contributing Author

Stephen G. Persson

Captain

Cambridge Fire Department
Cambridge, Massachusetts

Captain Steve Persson is a thirty-five year veteran of the Cambridge Fire Department.

He was promoted to Captain in 1986 and assumed the role of Department Training Officer in 1991. He has retained this position for the past 16 years. The Training Division is responsible for the every day training of the 278 member department including basic and advance firefighting skills, emergency medical services, driver training, hazardous materials, and homeland security programs. Initial recruit instruction is also conducted after graduation from the Massachusetts Firefighting Academy.

Captain Persson has planned, coordinated, and lent support to several mass casualty exercises, hazardous materials, and weapons of mass destruction training programs in which the Cambridge Fire Department, mutual aid departments and the public and private sectors have participated. Captain Persson was instrumental in implementing an Incident Management System within the department that has been in effect for the past 14 years. Captain Persson holds an AS degree in Fire Science Technology, has attended the University of Massachusetts Institute of Government Services, is a certified Instructor I, and a state certified Harzardous Materials Technician. Captain Persson has been appointed by the Governor as a member of the Massachusetts State Fire Training Council and represents the interests of the Massachusetts Institute of Fire Department Instructors. He has worked at the National Fire Protection Association in Quincy, Massachusetts in the Public Fire Protection Division as a Fire Service Research Assistant, author, technical assistant, and consultant. In addition to his duties on the Cambridge, Massachusetts Fire Department, he is the Fire Chief and Town Fire Warden in the town of Frye Island, Maine.

Preface

Engine company personnel are a key part of firefighting operations on the fireground. This book emphasizes that fire fighters performing engine company tasks must be properly trained, possess the proper equipment, and be adequately staffed. **Engine Company Fireground Operations** covers the basic objectives of engine company work including the proper supply and use of water to fight fires. This book also emphasizes that the engine company should be focused on three major tactical priorities: life safety, extinguishment, and property conservation. Some other areas of importance covered in this book include protecting exposures, confining the fire, and carrying out overhaul operations.

As many of these engine company tasks are carried out in the dangerous environment of flame and smoke, this book also stresses that it is essential for fire fighters to understand the nature of fire and the factors that affect its spread, including building construction, type of occupancy, and types of fuel available to the fire. This book also covers some of the recent improvements in firefighting equipment, which are designed either to supplement the use of water against a fire or to increase its efficiency.

Engine and ladder company operations are both critical to the success of a fireground operation, and these efforts must be coordinated in a safe and effective manner. Engine companies are responsible for providing water to fight a fire and the personnel to use it at the fireground. The examples used in this book do not include all the duties of an engine company, but rather illustrate some important points about engine company work.

Engine Company Fireground Operations provides the reader with a concise look at the issues facing many fire departments. This book, along with the companion book, **Ladder Company Fireground Operations**, aims to assist fire departments as they develop and conduct effective operations. These books provide overall basic fireground procedures required for effective firefighting activities.

Chapter 1 of **Engine Company Fireground Operations** introduces the typical duties of an engine company and how their efforts impact fireground operations. It also covers the nature of fire and the factors that affect its spread. Chapter 2 focuses on the combination of personnel performance and equipment which helps determine the outcome on the fireground. Chapter 3 discusses the issue of how engine companies are positioned at a fireground. This chapter also covers the importance of pre-incident planning, basic coverage responsibility, and the types of buildings that can pose coverage problems.

The issue of rescue, which is the first and highest priority at an incident, is discussed in Chapter 4. Chapters 5 through 7 cover the issues of water supply, initial attack lines, and backup lines. These chapters also address some of the hazards that should be avoided on the fireground. Exposure protection, which is essential to protect a building or area that has been subjected to radiant, convected heat, or direct flames, is addressed in Chapter 8.

The use of master stream appliances at the fireground is discussed in Chapter 9. Chapter 10 covers fire protection systems to mitigate the impact of a fire on a building and its occupants. Chapter 11 covers the importance of overhaul activities, which are dangerous, but necessary to ensure that all fire is extinguished and the area is safe.

This book provides a comprehensive overview of engine company operations at the fireground. The book also identifies some of the tools, procedures, and techniques that can be used for enhancing fire department operations to make the operations run as efficiently as possible.

Captain Stephen Persson
Cambridge Fire Department
Cambridge, Massachusetts

Introduction

Chapter 1

Learning Objectives

- Understand the three major tactical priorities on the fire ground: life safety, extinguishment, and property conservation.

- Understand the theory of fire spread and the four avenues that heat can travel.

- Recognize the potential for the sudden ignition of combustibles known as flashover.

- Examine the possibility of accumulated gases igniting into a rapidly spreading fire or violent explosion.

- Assess the primary responsibilities of an engine company on the fire ground.

The engine company is the basic unit of a fire department. In fact, many fire departments consist of only engine companies. The reason is simple: The engine company provides the primary firefighting agent—water—and the personnel to apply it properly on the fire ground.

In the last several years, a number of technical improvements in firefighting equipment have been made. Personal protective equipment and a self-contained breathing apparatus help the fire fighter advance to the seat of the fire by providing a measure of protection from heat, gases, and physical injury. Technical advances made on equipment such as nozzles and associated appliances, hose, including large-diameter hose used for supply hose lines and for providing water to master stream appliances, a wide variety of both hand and power tools and equipment, and electronic devices such as the thermal imaging camera have allowed for a more effective fire attack. New, efficient extinguishing agents are also available. For the most part, these improvements are designed either to supplement the use of water against a fire or to increase its efficiency. In no way do they detract from the importance of water in firefighting operations.

To supply and use water properly, the fire fighter must have considerable skill and knowledge, along with a certain amount of brawn and the ability to withstand physical and mental stress. The engine company must be well trained. For fire fighters to become proficient at any task on the fire ground, they must train. Fire departments do not respond to working fires on a daily basis; therefore, fire fighters need to hone their skills by continually training. NFPA 1410, *Standard on Training for Initial Emergency Scene Operations*, is a training standard designed to provide fire departments with an objective method of measuring performance for initial fire suppression and rescue procedures using available personnel and equipment.

The standard specifies basic evolutions that can be adapted to local conditions and serves as a standard mechanism for the evaluation of minimum acceptable performance during training for initial fire suppression and rescue activities. The engine company should be equipped and adequately staffed to carry out all of the objectives of a firefighting operation. These objectives, which have evolved over years of firefighting experience, form the basis for any fire attack plan. The three major tactical priorities on the fire ground are as follows:

- Life safety
- Extinguishment
- Property conservation

Key Points

The three major tactical priorities on the fire ground are life safety, extinguishment, and property conservation.

Other firefighting tasks may be intertwined among the three tactical priorities, such as water supply, protecting exposures, confining the fire, and overhaul operations. The three tactical priorities are listed in the order that they should be carried out. This does not mean that the first priority must be accomplished before another is started. Fire-ground priorities may be occurring simultaneously to ensure the protection and safety of both occupants and fire fighters.

All of these objectives can be carried out in a hazardous atmosphere of flame and smoke. Thus, fire fighters must understand the nature of fire and the factors that affect its spread, including building construction, type of occupancy, and types of fuel available to the fire.

Fire Spread

Oxygen, fuel, and heat are required to start and sustain the combustion process. These form the three sides of the familiar fire triangle **Figure 1-1**. A more advanced concept of combustion includes a fourth element, a chemical chain reaction phase, to form a **fire tetrahedron**. The chemistry of fire is not covered in this text, nor are the technical aspects of the support of combustion. Fire fighters making an attack are confronted with the problem after the fact. Thus, the discussions in this text are directed toward understanding how the fire advances through a building and how it extends to exposures. These characteristics of fire affect the firefighting operation.

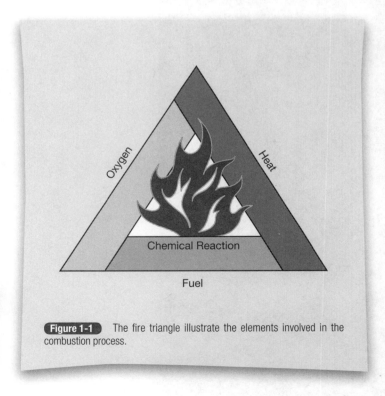

Figure 1-1   The fire triangle illustrate the elements involved in the combustion process.

Key Points

There are four ways heat travels: convection, radiation, conduction, and direct flame contact.

In structural fire situations, the fuel and oxygen required to sustain a fire are generally in plentiful supply. The fire usually starts out small, and if attacked early enough, it could easily be confined to the vicinity of its origin. When the fire burns unchecked, heat production increases. As the original fuel sources are consumed, the fire travels to new fuel sources in uninvolved parts of the building and in exposures. There are four ways heat travels:

• Convection
• Radiation
• Conduction
• Direct flame contact

Convection

Convection is the travel of heat through the motion of heated matter—that is, through the motion of smoke, hot air, heated gases that the fire produced, and flying embers. When it is confined, for example, within a structure, convected heat moves in predictable patterns. The fire produces gases that are lighter than air and that rise toward the top of the building **Figure 1-2**. Heated air, now buoyant, also rises, as does the smoke that

Figure 1-2 Convection carries hot air, smoke, gases, and embers upward through available vertical channels.

hence chimney of effect or reaction

combustion produced. As these heated combustion products rise, cool air takes their place; the cool air is heated in turn and then also rises to the highest point that it can reach **Figure 1-3**. As the hot air and gases rise away from the fire, they begin to cool; as they do, they drop down to be reheated and rise

?

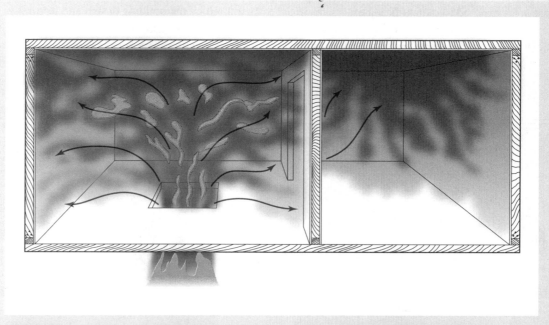

Figure 1-3 When vertical travel is blocked, convection carries hot air, smoke, gases, and embers horizontally.

again. This is the **convection cycle**. Within a building, it will fill first the upper parts and then lower parts of the area above the fire.

It is easy to see how this method of heat and fire transmission creates a need for rescue operations and checks for fire spread in the building. In addition, convection is the main reason for ventilation requirements in fire department operations. Modern building codes are aimed at limiting fire spread by convection through fire resistant construction, which is intended to confine convection currents to one floor or to a small area of a floor.

Radiation

Radiation is the travel of heat through space; no material substance is required. Pure heat travels from the fire area in the same way as light (i.e., in straight lines). It is unaffected by wind, and unless blocked, it is radiated evenly in all directions **Figure 1-4**. After the fire has built to sizable proportions, radiation is the greatest cause of exposure fires, spreading fire rapidly from structure to structure or through storage areas. Intense radiant heat drives fire fighters back from normal approach distances, necessitating the use of master streams to bring water to bear on the fire and exposed structures or stored materials.

In combination with convected heat, radiation creates the most severe area of exposure; this area must be protected first. Just because wind does not affect radiation, however, fire fighters cannot ignore the windward side of the fire. Fire departments have been caught short too often when all of their efforts are being directed to the area hit by both radiant and convected heat. After this most dangerous area has been covered, attention must be given to those areas exposed to radiant heat alone.

Within the confines of a building, radiant heat quickly raises the temperature of air and combustible material both near and if the building layout permits at quite some distance from the

fire. Flashover may occur long before the flames actually contact the fuel in a given area.

Proper ventilation is of little help against concentrations of radiant heat. Venting will remove the smoke, hot air, and heated gases, thereby lessening the chance of rapid spread and flashover, but the radiant heat remains and must be counteracted by fire fighters through proper application and adequate amounts of water on the seat of the fire.

Conduction

Conduction is the travel of heat through a solid body **Figure 1-5**. Although it is normally the least of the problems at a fire, the chance of fire travel by conduction should not be overlooked. Conduction can take heat through walls and floors by way of pipes, metal girders, and joists and can cause heat to pass through solid masonry walls.

If the spread of fire by conduction occurs at all, the time involved will depend on the amount of heat and fire being applied to a structural member or wall. In any case, when fire fighters find that fire has been in contact with such parts of the building, they must check thoroughly to be sure that fire has not traveled through them to other areas. They must also be aware that heat can be conducted down, as well as in other directions, if permitted by the building design and features in the fire area.

Conduction also can be dangerous to fire fighters. Certain types of structures have steel building and roof supports that are completely open to fires. Heat spreading through these supports

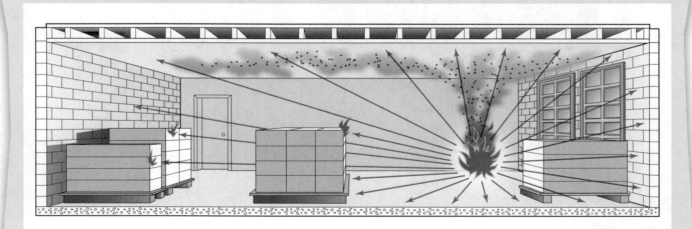

Figure 1-4   Heat is radiated evenly in all directions from the fire.

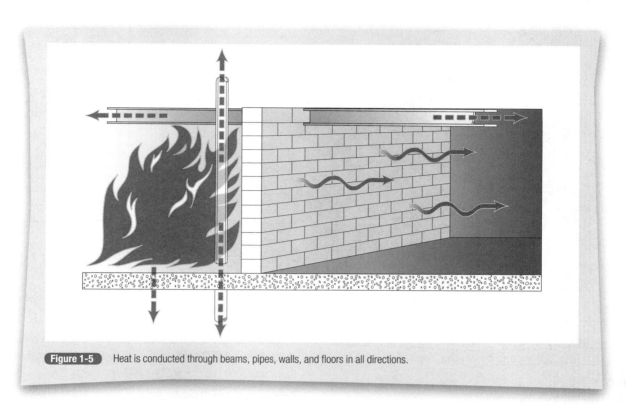

Figure 1-5 Heat is conducted through beams, pipes, walls, and floors in all directions.

raises their temperature and may cause them to warp and fail, possibly causing the walls and roof to collapse **Figure 1-6**. A 100-foot section of steel beam will expand 9 inches when heated to 1000°F.

In many cases, the action of hose streams stops conduction without a fire fighter being aware of it. The cooling water removes heat from the structural members, walls, and floors involved in conduction. Heat traveling by conduction may not be seen or felt by fire fighters in the normal course of operations at a fire. Fire fighter must be observant, however, and check for the spread of fire by conduction when there are indications that this action could take place. A fire started by a heated pipe or metal

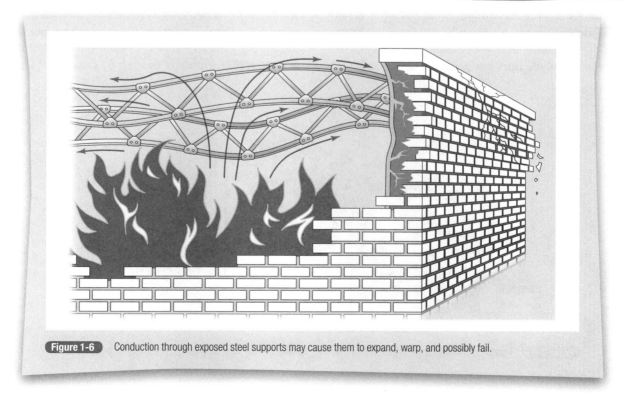

Figure 1-6 Conduction through exposed steel supports may cause them to expand, warp, and possibly fail.

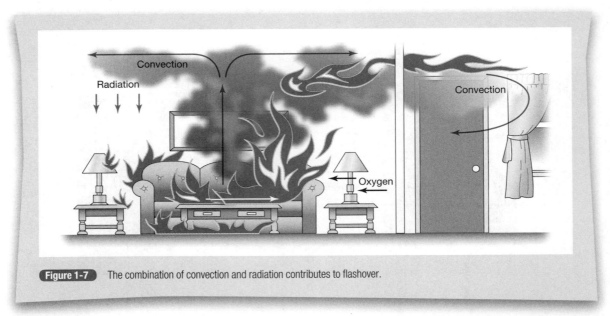

Figure 1-7   The combination of convection and radiation contributes to flashover.

structural member in an area remote from the main fire is one indication that heat is traveling by conduction.

Flashover

Flashover is the ignition of combustibles in an area heated by convection, radiation, or a combination of the two. The action may be one of sudden ignition in a particular location followed by rapid spread or one of a "flash" of the entire area. The latter action is more likely to happen in an open area.

Convection may cause rollover at the top of a structure because of the hot products of combustion igniting materials at that level. Rollover is a condition where unburned products of combustion from a fire accumulate at the ceiling level in a sufficient concentration that they ignite momentarily. Radiation may contribute to flashover in areas that do not block heat travel; however, radiation alone does not generally cause flashover.

Radiation usually acts in combination with convection as a cause of flashover **Figure 1-7**. Flashover may cause ignition at some distance from the fire **Figure 1-8**.

Smoldering Fire, Backdraft

During the early stages of a fire, the products of combustion may fill the building until the fire is almost starved for oxygen, at which point it will begin to smolder. The more incomplete the combustion, the more carbon monoxide produced. Carbon monoxide is a colorless, odorless, poisonous gas. Because it is also explosive and flammable, a dangerous situation exists. Two of the elements necessary to produce fire—heat and fuel—are contained within the structure; only oxygen needs to be added. If oxygen is allowed to enter the structure, the accumulated gases will ignite into a rapidly spreading fire or a violent explosion. This is known as a backdraft. Fortunately, fire fighters can control this situation

Figure 1-8   Flashover, heated gases, and radiant heat traveling through a building may cause ignition at some distance from the fire.

effectively through proper ventilation and fire attack procedures (see Chapter 6 for more information on backdraft).

Engine Company Operations

First and foremost, a fire department must operate using a structured **Incident Management System (IMS)** on the fire ground. It is required by law. The IMS is an organized system of roles, responsibilities, and standard operating guidelines used to manage emergency operations. Engine companies must work within an overall strategic plan developed by the incident commander. Each engine company must work within the system and follow the plan by performing tasks assigned by the incident commander ensuring that fire fighter safety is maintained. The days of freelancing on the fire ground are over, and all fire fighters working at an incident must be accountable for their actions. Standard operating guidelines should be in place, and engine companies should adhere to these procedures during fire ground operations. The primary responsibilities of an engine company are life safety and applying water onto the fire. During an interior (offensive) operation, an engine company must be able to

- Place a hose line between the fire and victims or rescuers to protect them and provide a safe evacuation route from the building
- Maintain the integrity of interior stairways for entrance and egress
- Provide a primary search while advancing a hose line
- Advance a hose line to the seat of the fire for extinguishment

During an exterior (defensive) operation, an engine company must be able to

- Set up and operate master stream devices to protect exposures
- Extinguish the main body of fire

Engine company fire apparatus and equipment have been designed to allow fire fighters to function effectively and quickly. Through training and experience, personnel must acquire the knowledge, skill, and judgment in performing the following nine basic firefighting tasks of an engine company:

- Performing search and rescue operations
- Establishing a water supply
- Use of initial attack lines
- Use of backup lines
- Protecting exposures
- Use of master stream appliances
- Tactical use of protective systems
- Property conservation
- Performing overhaul operations

It is not expected that one engine company will perform every one of these operations at every fire, nor are the operations necessarily to be carried out in the order given here. Just as fire situations vary, what an engine company needs to accomplish at an incident will also vary. The fire apparatus should be properly equipped with both personnel and equipment, however, and fire fighters should be ready to carry out any task with particular attention focused on

the primary search. Each operation or task listed previously here is discussed in some detail in the chapters that follow.

In addition to the basic operations that are directly related to control of the fire, engine company personnel must also be proficient in property conservation operations. Even though property conservation has no bearing on the spread of the fire, it is very important in terms of total fire loss, including damage by water. Proper sizeup will determine how you prolong your actions—along with the location and intensity of the fire.

Knowledge of an engine company's district and information gathered from building inspections and preincident planning are of great help in operating efficiently at the fire ground. This information is also useful in developing and improving on standard operating guidelines. Engine company personnel should have a good working knowledge of their first-alarm district (territory), especially with regard to water supply, including hydrant locations and/or static water sources, life safety hazards, building construction, exposure hazards, locations particularly dangerous to fire fighters, target hazards, and other special conditions that affect firefighting operations. Perhaps it is impossible to learn everything about the company's response area, however, unusual situations should be carefully examined and analyzed, and special procedures should be developed when necessary.

These background data are the initial sizeup information that can be most useful when the engine company arrives at a working fire and assumes command. In addition, factors such as the time of day, weather conditions, and life safety considerations must also be considered. With this information, the incident commander will continue the sizeup from what he or she observes and from information received from other fire fighters and civilian personnel until command is transferred. Sizeup is a continuing procedure, resulting in operational changes to match changes in the fire situation. Because it affects engine company operations, it is mentioned from time to time in this book rather than being covered in a single chapter. **Sizeup** is the basis on which engine company operations are carried out, and thus, it is implicit in almost every sentence in the book.

Wrap-Up

Chief Concepts

- Fire fighters must understand the nature of fire and the factors that affect its spread—that is, convection, radiation, conduction, and direct flame contact.
- This understanding assists them in successfully carrying out three tactical priorities—life safety, extinguishment, and property conservation.
- Engine companies, as well as every other fire fighter at an incident, work within an IMS.
- This system must have an overall strategic plan.
- The incident commander, whether the first arriving company officer or the chief of department, is responsible for this strategic plan.
- Engine companies are responsible for performing life safety operations, establishing a water supply, and advancing and operating hose lines for both offensive and defensive modes of operation.
- Engine companies must be well trained, adequately staffed, and supplied with modern fire apparatus and equipment enabling them to perform their assigned tasks in an efficient and safe manner.

Key Terms

Backdraft: When oxygen enters a structure that is filled with the products of combustion and contains heat and fuel, the accumulated gases may ignite into a rapidly spreading fire or a violent explosion.

Conduction: The travel of heat through a solid body.

Convection: The travel of heat through the motion of heated matter.

Convection cycle: Heat transfer by circulation with a medium such as a gas or a liquid.

Fire tetrahedron: A geometric shape used to depict the four components required for a fire to occur: fuel, oxygen, heat, and chemical chain reactions.

Flashover: Ignition of combustibles in an area heated by convection, radiation, or a combination of the two.

Incident Management System (IMS): An organized system of roles, responsibilities, and standard operating guidelines used to manage emergency operations.

Radiation: The travel of heat through space; no material substance is required.

Sizeup: Basis on which engine company operations are carried out.

1. The _EngineCompany_ is the basic unit of a fire department.
   a. Engine company
   b. Ladder company
   c. Rescue company
   d. Depending on the fire department, either the engine or ladder company could be considered the basic unit

2. New equipment such as class A foam and large diameter hose _____.
   a. Make water less important in terms of firefighting
   b. Are of little value in terms of actual suppression
   c. Permit fire fighters to implement a defensive strategy that is nearly as effective as an offensive strategy
   d. Supplement or increase the efficiency of water

3. When applying the three tactical priorities of life safety, extinguishment, and property conservation:
   a. All life safety activities must be completed before extinguishment or property conservation, but extinguishment and property conservation activities can be accomplished simultaneously.
   b. All life safety activities must be completed before extinguishment, and all extinguishment activities must be completed before beginning any property conservation tasks.
   c. The three tactical priorities are merely guidelines and can be completed in any order.
   d. Property conservation tasks can be started before the completion of life safety and extinguishment activities.

4. _____ heat carries hot air, smoke, and gases from a lower floor to an upper floor.
   a. Conducted
   b. Convected
   c. Particulate
   d. Radiant

5. Heat that travels through space in a straight line to ignite an exposure would be an example of fire spread via _____ heat.
   a. Conducted
   b. Convected
   c. Particulate
   d. Radiant

6. Heat that travels via a metal beam through a fire wall to ignite combustibles on the other side of the fire wall would be an example of fire spread by _____ heat.
   a. Conducted
   b. Convected
   c. Particulate
   d. Radiant

7. Flashover generally results from
   a. Convected heating
   b. Conducted heating
   c. Radiant heating
   d. Convected heating, radiant heating, or a combination of the two

8. A room is prime for a backdraft when the _____ side of the fire tetrahedron is absent.
   a. Chemical chain reaction
   b. Fuel
   c. Heat
   d. Oxygen

9. Which of the following is *not* a basic engine company task:
   a. Performing search and rescue operations
   b. Performing ventilation operations
   c. Performing overhaul operations
   d. Property conservation

Equipment and Initial Hose Operations

Chapter 2

Learning Objectives

- Know the equipment on an engine company that is required for a safe and efficient operation on the fire ground.

- Define the four basic hose lays used by a one-piece (single-pumper) engine company.

Company officers and fire fighters, no matter how well trained, cannot perform efficiently without proper equipment and adequate personnel. The combination of personnel performance, training equipment familiarity, and equipment determines the outcome on the fire ground. Because engine company personnel will be called on to engage in combating many different types of fires, their equipment must be chosen for its quality and dependability to assure adequate performance while covering a wide range of fire-ground situations.

This chapter lists and discusses several pieces of equipment required for an efficient engine company operation. As sufficient staffing levels are often a problematic situation in both paid and volunteer departments, equipment should be chosen that allows a limited number of fire fighters to carry out fire-ground operations safely and efficiently.

Standard equipment requirements for an engine company are basically alike in urban, suburban, and rural areas. Modifications to the list will most likely occur in the pumping capacity of the pump, the size of the water tank, tank configuration, and any specialized tools and equipment that may be needed in a specific city, town, or fire district. It is the responsibility of the authority having jurisdiction (AHJ) to ensure that the fire apparatus and the equipment carried on such apparatus is compliant, adequate, and functional for the task.

It is not enough to be well trained and adequately staffed and equipped. Time is the ally of the fire, not of the fire fighter. Fire doesn't slow down so that fire fighters can play catch up. An engine company must deploy equipment quickly at the scene of a fire. The first few minutes at the scene of a working fire could mean the difference between a successful operation and a failed one. A wise fire-ground adage is that as the first attack line goes so goes the rest of the fire. An initial task of an engine company is to provide a water supply to the fire ground. The second section of this chapter discusses a number of initial supply hose lays. Departmental standard operating guidelines should dictate the procedures used to deliver adequately an uninterrupted supply of water to the fire ground. Consideration should be given to the types of apparatus responding, the arrival time of additional companies, existing water sources and their capacity, hose and appliances, available personnel, the size and type of building construction, and the accessibility of mutual aid.

Engine Company Equipment

NFPA 1901, *Standard for Automotive Fire Apparatus*, is a valuable document that defines the requirements for new automotive

fire apparatus designed to be used under emergency conditions to transport personnel and equipment and to support the suppression of fires and mitigation of hazardous situations.

The AHJ should be familiar with documents that will allow them to make informed decisions before purchasing fire apparatus and equipment. In this way, a fire department will not be surprised when equipment arrives and does not meet their requirements or standards. Apparatus manufacturers can be of assistance to your department.

The engine itself should be equipped with a pump having a rated pumping capacity of no less than 750 to 1,000 gallons per minute and a water tank that carries at least 750 to 1,000 gallons of water. These are absolute minimums for the general operation of engine companies. The capacities should be increased on the basis of a fire department's knowledge of its needs and standard operating guidelines. The fire department should choose a water tank of a size that best supports efficient and effective fire-ground operations. For example, a rural engine company might find that a minimum 1,000-gallon water tank is required to maintain an initial attack during the length of time needed to set up its pumper. If so, that engine company should be equipped with a 1,000-gallon water tank.

NFPA 1901 states that the requirements of service in different communities might necessitate additions to the equipment required. The operational objective is to arrive at the scene of the emergency with the necessary equipment for immediate life safety operations and emergency control. NFPA 1901 recommends that **pumper fire apparatus** carry the partial list of equipment described in the following sections. Remember, a pumper is not a tanker.

Hose Storage

NFPA 1901 requires a minimum hose storage area of 30 ft^3 for 2½-inch or larger fire hose and two areas, each a minimum of 3.5 ft^3 to accommodate 1½-inch or larger preconnected hose lines. A **divided hose bed** is one that is separated into two supply hose compartments running the length of the hose bed.

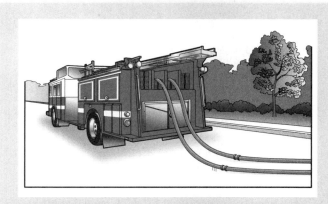

Figure 2-1 A divided hose bed allows two supply lines to be laid simultaneously.

This arrangement permits two separate hose lines to be laid simultaneously **Figure 2-1**. It also allows for two different hose setups for two different types of hose lays. For example, one side of the hose bed may be set up for a reverse lay and the other for a forward or straight lay **Figure 2-2**. This setup allows for flexibility of operation at the fire ground. A divided hose bed can be used with a 2½-inch, 3-inch, or large-diameter hose (LDH) of 4- and 5-inch diameters.

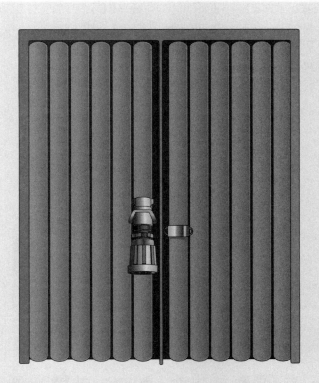

Figure 2-2 In the divided hose bed shown, the section on the right is set up for a forward lay (water to fire), and the section on the left is set up for a reverse lay (fire to water).

Key Points

Hose storage areas should be arranged so that the configuration will best support the operational procedures following standard operating guidelines.

Hose storage areas should be arranged so that the configuration will best support the operational procedures following standard operating guidelines. There are many variations as to how supply and attack hose lines are carried on fire apparatus.

Fire Hose

NFPA 1901 requires a minimum of 400 ft of 1½-, 1¾-, or 2-inch fire hose to be carried on pumper fire apparatus. Although 1½- and 2-inch hose are used by fire departments for attack lines, 1¾-inch hose has become the most widely used within the fire service for this purpose. The 1½- or 1¾-inch hose, with 1½-inch couplings, can be quickly advanced for rescue, interior exposure coverage, and direct fire attack **Figure 2-3**. The size and capacity of the fire hose allow fast movement and sufficient

Key Points

NFPA 1901 requires a minimum of 400 ft of 1½-, 1¾-, or 2-inch fire hose to be carried on pumper fire apparatus.

Figure 2-3 The highly mobile 1¾-inch lines can be quickly advanced for rescue, interior exposure coverage, and direct attack.

delivery of water for most interior fires; however, 1½- and 1¾-inch hose lines cannot be used on every fire in spite of their mobility. When the size or intensity of the fire dictates the use of larger hose lines, with greater reach and capacity, then they must be placed in service. The basic rule of thumb is this: If you have a big fire, you operate a big line.

Wherever possible, the 1½- or 1¾-inch hose lines should be preconnected to pump outlets. **Crosslays**, or transverse hose beds, are usually located behind the crew compartment. The configuration of the hose bed allows attack lines to be deployed from either side of the pumper. These are connected in the hose bed with a swivel valve. The hose is loaded so that it can be stretched quickly upon arrival at the scene. Fire fighters, with the help of the driver/operator, should be able to get this attack line in position, changed, and in operation in a minimal amount of time.

If this is not feasible, separate hose beds can be provided, with the 1½- or 1¾-inch hose lines loaded so that the nozzles are on top. The nozzle and hose line can then be advanced efficiently to the firefighting position. When the crew at a nozzle is in position, the hose can be broken at a convenient coupling and attached to the pump discharge with a 2½- to 1½-inch reducer **Figure 2-4** . Although this method works, modern fire apparatus must be equipped with crosslays for preconnected hose line operations.

Preconnected hose lines of 1½- or 1¾-inch should not exceed 250 ft in length because of excessive friction losses in

longer lays. If these hose lines need to be extended a further distance, they should be connected to larger diameter hose lines. The most common lengths of 1½- or 1¾-inch hose on preconnected lays are 150 and 200 ft. A gated wye is often used and is very advantageous when attached to the end of the larger diameter line.

2½-Inch Fire Hose

NFPA 1901 requires a minimum of 800 ft of 2½-inch or larger fire hose. The 2½-inch hose line is recommended for fires that cannot be controlled by smaller hose lines **Figure 2-5** . Although the 2½-inch line is bulkier and more difficult to maneuver and operate, its water delivery capacity is absolutely necessary for attacking large, intense fires. Probably because of a lack of proper staffing, there is a strong tendency to use the smaller hose lines for the initial attack. Some fire departments assume that the 1¾- or 2-inch hose line is the universal remedy for controlling all fires. Nothing is further from the truth. In fact, since the introduction of this smaller diameter fire hose, many fire departments have eliminated 2½-inch fire hose. This is a big mistake. If the fire is obviously too large for smaller attack lines, they should be left on the pumper.

Figure 2-4   A reducer is used to connect hose couplings of different diameters.

Figure 2-5   Fires that have gained some headway should be attacked with 2½-inch hose lines.

For efficient operation, a 2½-inch preconnected hose line should be set up in a separate compartment on the pumper, and a spray nozzle or solid-stream nozzle should be attached. Because of the reduced nozzle pressure and lower nozzle reaction, a line with a solid-stream nozzle will be easier to maneuver and operate inside the fire building. The nozzle should be a leader line type. This tip is equipped with a screw-down thread at the end. If needed, a smaller diameter hose with 1½-inch couplings can be attached to the 2½-inch hose line using the screw-down feature on the tip. In this way, a smaller diameter hose line could be advanced a greater distance than 200 ft from the pumper. The attack line could be used if this smaller diameter line was adequate for the current task. A preconnected 2½-inch hose line with a solid-bore nozzle and a smooth-bore tip leader line can be used with smaller diameter hose attached to the screw-down tip. Many departments also have an adapter above the shutoff on spray nozzles that allows the tip to be removed and a smaller line extended. The advantage of either of these lines is that the 1½-inch connection is ahead of the shutoff; thus, the 2½-inch line can remain charged while the smaller hand line is extended. (Both types of nozzles are commonly available with the break-apart or leader line to allow a smaller line to be added or used for overhaul.)

Another example in which a 2½-inch hose line would be beneficial is if the 2½-inch hose line is used when there is a long stretch from the pumper to the fire. If 1½- or 1¾-inch hose lines are sufficient for fighting the fire, the 2½-inch hose line can be stretched from the pumper toward the fire. Two 1½- or 1¾-inch hose lines can be advanced from a double-gated reducing leader wye to combat the fire.

As with the 1½- and 1¾-inch hose line, the preconnected 2½-inch hose line can be placed into service with a minimum of delay. If discharge outlets are not available for a 2½-inch preconnected hose line, a separate hose compartment can be set up to hold 150 to 250 ft of 2½-inch hose. This hose line can be quickly removed and hooked up to a standard outlet when needed.

Supply Hose

A **supply hose** is designed for the movement of water between a pressurized water source and a pump that is supplying attack lines. A supply hose is an LDH of 3½ inches or larger. LDHs provide the movement of large amounts of water to the fire ground from a water source with less friction loss and fewer fire fighters to establish a permanent water supply. Most larger diameter hoses are 4 or 5 inches.

In areas without a municipal water system, an LDH has become a temporary above-ground water main from the source to the fire ground. An LDH offers many advantages, especially with a limited number of personnel. The main advantage of an LDH is not only an increase in flow area, but also a corresponding reduction of friction loss compared with a common 2½-inch hose. NFPA 1142, *Standard on Water Supplies for Suburban and Rural Fire Fighting*, states that where delivery rates exceed 500 gallons per minute and water is moved long distances, an LDH provides a most efficient means of minimizing friction loss and developing the full potential of both water supplies and pumping capacities.

Another major advantage of an LDH is increased productivity, as the same number of personnel and apparatus can use an LDH to move more water over longer distances. Extended horizontally, a 4-inch LDH flows almost 3.5 times the capacity of 2½-inch hose of equal length; therefore, an LDH will allow fire fighters to get the operation going in a faster, more reliable fashion with a larger flow, less friction loss, and less physical stress on personnel.

Nozzles

The purpose of a nozzle is to shape the stream and convert pressure energy to velocity energy. Many specialized nozzles exist; however, they can be classified in two ways: solid-stream and spray or fog nozzles. The NFPA requires the following nozzles to be carried on the pumper apparatus:

- One combination spray nozzle, 200 gpm minimum
- Two combination spray nozzles, 95 gpm minimum
- One playpipe, with shutoff and 1-, 1⅛-, and 1¼-inch tips

Solid-Stream Nozzles

Solid-stream nozzles are classified according to the nozzle diameter . Nozzles with tips up to 1⅛ inch or perhaps 1¼ inch are generally considered for use on hand lines (i.e., those

(a) Solid-stream nozzle   (b) Spray (or fog) nozzle

Figure 2-6   Solid stream and spray are two types of nozzles.

normally held in the hands). A 1¼ inch tip is the breaking point for handlines and streams at 50 psi versus 80 psi.

The 1⅛-inch nozzle used with 2½-inch hose produces the so-called standard fire department stream of 250 gallons per minute at about 45 psi nozzle pressure. Many fire departments use hand-held hose lines with flows exceeding 250 gallons per minute. Department members must train and be thoroughly familiar with hose lines flowing maximum gallons per minute before using these high-flow nozzles on the fire ground. Nozzles with tips larger than 1¼ inch have such large reaction force that they must be mechanically restrained. Tips from 1¼ to 2 inches are usually used on master stream appliances, such as monitors, deluge sets, or deck guns.

Solid streams are useful where extreme range is desired and where penetrating capabilities are needed, such as where thermal degradation of spray streams prevents proper penetration, or for force of the stream **Figure 2-7** . At winds above 30 miles per hour, you will need to use a straight tip for penetration. Solid streams do not have as good heat-transfer characteristics, as spray nozzle streams and consequently are not as effective in absorbing heat.

Spray Nozzles

Spray nozzles are also called fog nozzles and produce varying degrees of water spray (Figure 2-6). They may have a fixed-spray angle or may be adjustable from almost a straight stream to a very wide-angle spray. Many spray nozzles have predetermined pattern settings, usually at straight stream, 30-, 60-, and 90-degree spray angles; however, most can also be set at intermediate angles.

The designated flow from a spray nozzle is usually rated at 100-psi nozzle pressure, although there are now low-pressure spray nozzles that operate at lower pressures. Some spray nozzles have different flows at various angles. Other spray nozzles are constant gallonage—that is, they have the same flow at all spray angles. In most interior structure operations, the fog patter is no greater than 30 degrees.

Spray nozzles set at a fog pattern are effective in more quickly absorbing heat than those set at a straight stream pattern (Figure 2-7). When using spray nozzles for interior firefighting operations, extreme care must be taken not to drive fire and the products of combustion into uninvolved areas of the building.

Key Points

Solid streams are useful where extreme range is desired and where penetrating capabilities are needed, for example, where thermal degradation of spray streams prevents proper penetration.

Key Points

Spray nozzles set at a fog pattern are effective at more quickly absorbing heat than those set at a straight stream pattern. The rapid absorption of heat will quickly spread the steam produced throughout the building, endangering occupants and fire fighters.

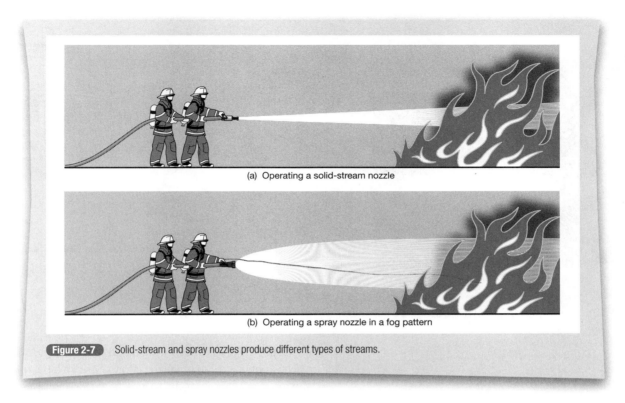

(a) Operating a solid-stream nozzle

(b) Operating a spray nozzle in a fog pattern

Figure 2-7 Solid-stream and spray nozzles produce different types of streams.

Fire should be attacked from the unburned side if possible, pushing fire, heat, and the products of combustion back into areas already affected by the fire. Chapter 6 contains additional information on solid-stream and spray nozzles.

Master Stream Appliance

The **master stream appliance** is a portable unit that can be used either mounted on or detached from the pumper. It is capable of immediate application of a heavy stream. NFPA 1901 recommends a master stream appliance of 1,000 gpm minimum. Initially, the master stream appliance may be supplied from the water tank and used to help knock down the main body of fire, thereby allowing an attack with hose lines. Because the master stream appliance will soon empty the water tank, its continued operation requires a quick connection of the pump to the water system to maintain an uninterrupted supply of water.

Prepiped Master Stream Appliance

A **prepiped master stream appliance** can be put into operation as soon as the pumper is at the fire and the nozzle is positioned. A prepiped master stream appliance has a separate discharge pipe of adequate diameter that runs from the fire pump to the appliance. Master stream appliances on engine companies swivel 360 degrees, raise up and down, and may have a telescoping feature. Opening a gated discharge outlet allows water to flow into the appliance. The appliance is easily detached from its base on the apparatus and can be operated from the ground supplied by one or more hose lines.

When the master stream appliance is not preconnected, one or two short sections (15 to 25 ft in length) of 2½- or 3-inch hose or an LDH should be rolled in doughnut rolls and connected to the appliance **Figure 2-8**. This allows the appliance to be connected to the pump's discharge gates and placed in operation rapidly.

Soft-Suction Hose

NFPA 1901 requires a minimum of 15 ft of soft-suction or 20 ft of hard-suction hose. The soft-suction hose allows a quick, efficient connection of the pump intake to a hydrant or other pressurized water source. Soft-suction hose is generally available in sizes from 2½ to 6 inches, with 4 and 5 inches being the most popular sizes. A hard-suction hose is primarily used for drafting water from a static water source such as a lake, pond, or swimming pool.

Key Points

The master stream appliance is a portable unit that can be either mounted on or detached from the pumper for the immediate application of a heavy stream.

Key Points

A prepiped master stream appliance can be put into operation as soon as the pumper is at the fire and the nozzle is positioned.

2½ in. or 3 in. doughnut roll

2½ in. or 3 in. doughnut roll

Figure 2-8 Two short sections of hose can be connected to the master stream appliance for quick hookup to the pump.

Figure 2-9 Front or rear intakes in combination with an attached soft-suction hose permit good positioning and quick hydrant hookup.

It is also used during tanker shuttle operations for drafting, or siphoning, water from a portable water tank. A hard-suction hose is generally available in sizes from 2½ to 6 inches. Pumpers will carry a minimum of two 10-foot section of hard sleeves (hard suction).

If one end of the soft-suction hose is preconnected to the pump intake, one fire fighter can quickly stretch the unattached end, couple it to the hydrant, and charge the supply hose line. A side or rear intake may be used, but normally the soft-suction hose is connected to the front intake. Drafting from a static water source is most efficient from a side intake because of the length of pipe on the front of the pump.

Pump Intake Connections

The NFPA 1901 requires that a pump have at least the number of intakes required to match an arrangement for the rated capacity of the pump, and the required intakes must be at least equal in size to the size of the suction lines for that arrangement.

Pump intake connections are used in conjunction with soft-suction hose for hydrant operations and with hard-suction hose for drafting evolutions. Front or rear intakes permit better

positioning of the pumper close to the curb than do side intakes. The front intake is particularly effective on cab forward apparatus when connecting directly to a hydrant **Figure 2-9**.

For drafting, a front intake permits easier maneuvering and better positioning of the pumper. On cab forward apparatus, a front intake can be set close to the water while the front wheels remain on firm ground **Figure 2-10**.

Figure 2-10 A front intake permits easy and efficient drafting operations.

Key Points

The soft-suction hose allows a quick, efficient connection of the pump intake to a hydrant.

Key Points

Pump intake connections are used in conjunction with soft-suction hose for hydrant operations, and with hard-suction hose for drafting evolutions.

Key Points

An HAV allows a second pumper to be hooked up to the hydrant without shutting down the hydrant when a hydrant supply line is in operation supplying a pumper. It is especially useful if assistant pumpers have long response times or a delayed response.

2½-Inch Pump Intake Connections

A pumper located at a fire usually receives water from hydrants or other pumpers. To operate efficiently, the pump must receive the required amount of water through appropriately sized supply lines. Additional intakes are provided for this operation.

Most fire departments now use an LDH to supply pumpers on the fire ground. If needed, a 2½-inch pump intake connection would allow a pump to receive a supplemental supply line using a 2½-inch hose line or greater. In departments using 2½- or 3-inch hose to normally supply pumpers on the fire ground, the 2½-inch pump intake connection is well suited for supplemental purposes.

On existing units that have but one 2½-inch intake, a two-way siamese provides an alternate method of providing a supplemental water supply should the need arise. Two 2½-inch hose lines can be combined into one by using the siamese **Figure 2-11** .

Hydrant Assist Valve

The **hydrant assist valve (HAV)**, also know as four-way valve, comes in two sizes. The larger fits on the large hydrant discharge opening or steamer connection. The smaller HAV fits the 2½-inch hydrant

Key Points

Most fire departments now use an LDH to supply pumpers on the fire ground.

outlets. A HAV is attached to the end of a pumper's supply line. During an incident, the HAV is attached to the hydrant by the engine company. The pumper then proceeds to the fire ground, and the supply line is charged. If the volume of water from the hydrant needs to be increased, a second pumper can hook up to the HAV. Without shutting down the hydrant, the second pumper can hook up and increase the pressure and volume through the supply line to the pumper on the fire ground. An HAV can be a useful piece of equipment if used properly. It must be remembered that an HAV serves no purpose to a second pumper if the hydrant is unable to discharge a sufficient supply of water needed on the fire ground. If the pumper on the fire ground needs a supplemental water supply and the hydrant is unable to produce that supply, then the water should come from a different source with a sufficient volume.

Ball Valve

A **ball valve** has an internal plastic component shaped like a ball **Figure 2-12** . The ball has a hole through its center that allows water to flow through it when the valve is in the "open" position while no water is allowed to flow through it in the "closed" position.

Two ball valves placed on a hydrant allow an additional supply line to be laid without shutting down the hydrant **Figure 2-13** . A ball valve can be placed on the hydrant when the initial supply line is connected to the hydrant whether from the steamer connection or a 2½-inch discharge. A second supply line can be connected to the unused ball valve whenever it is

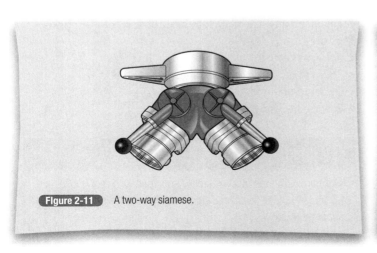

Figure 2-11   A two-way siamese.

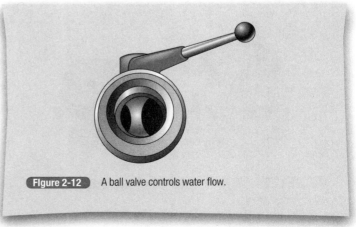

Figure 2-12   A ball valve controls water flow.

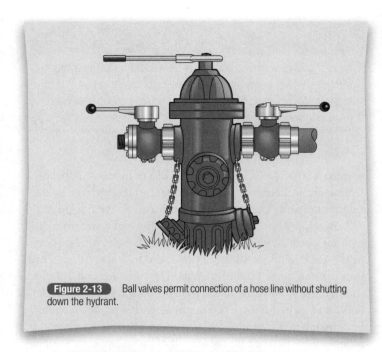

Figure 2-13 Ball valves permit connection of a hose line without shutting down the hydrant.

Key Points

Two ball valves placed on a hydrant allow an additional supply line to be laid without shutting down the hydrant.

needed; however, if a pumper has not been connected to the hydrant through a soft-suction hose, the hydrant must be shut down to complete the evolution.

Double Male and Double Female Fitting

Double male and double female fittings are used to connect two threaded connections of the same size and sex (Figure 2-14). They are needed when a pumper, set up for a forward lay, uses a reverse lay. To make the proper connections at the pumper or the nozzle, a double male and double female will be needed. NFPA

Figure 2-14 Double male and double female fittings connect two threaded connections of the same size and sex.

Key Points

Double male and double female fittings connect two threaded connections of the same size and sex. They are needed when a pumper that is set up for a forward lay uses a reverse lay.

1901 requires one double female 2½-inch adapter with National Standard Thread and one double male 2½-inch adapter with National Standard Thread. These are lightweight today.

Ground Ladders

The NFPA 1901 requires, at a minimum, that the following ladders be carried on the apparatus:

- One straight ladder equipped with roof hooks (12 or 14 feet in length)
- One extension ladder (24 foot minimum, but most pumpers run a 28-ft ladder)
- One attic ladder

Where there are no ladder trucks in service, pumpers should normally be equipped with a 35-ft extension ladder (Figure 2-15).

An engine company can use a ground or roof ladder to gain entry to upper stories for firefighting or rescue operations. Ladders carried on engine companies can also be used to operate hose lines in a fire building. There are a number of special circumstances in which ladders can be used, including ice rescues, bridging incidents, shoring, hoisting operations, and damming operations. Ladders may be carried on top or the side of the pumper and may use electric or hydraulic arms to lower the ladders to a safer ergonomic height and increase storage capacity.

One-Piece (Single-Pumper) Hose Lays

The great majority of fire departments operate one-piece engine companies; that is, each engine company has but one pumper. This section discusses the hose lays used by the one-piece engine company. There are four basic hose lays for supply hose. They are

- Forward lay using a charged supply line
- Forward lay using an uncharged supply line
- Direct to fire–no line laid (or split lay)
- Reverse lay using a charged line

Key Points

At a minimum, one straight ladder equipped with roof hooks, one extension ladder, and one attic ladder must be carried on the apparatus.

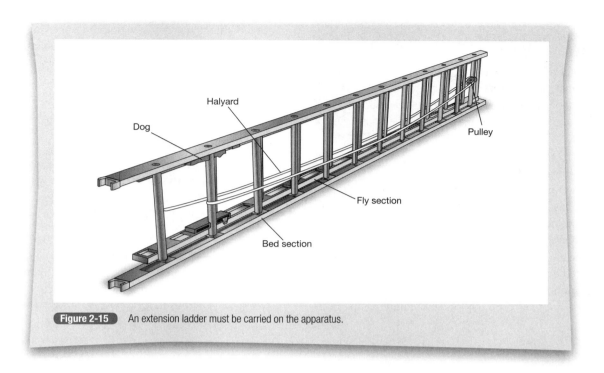

Figure 2-15   An extension ladder must be carried on the apparatus.

Key Points

The majority of fire departments operate one-piece engine companies; that is, each engine company has but one pumper.

Forward Lay Using a Charged Supply Line

Perhaps the most desirable hose lay is the forward lay from a hydrant to the fire, with the immediate charging of the supply line to the pumper **Figure 2-16**. This enables the engine company to function independently with an uninterrupted supply of water from the beginning of the operation. The pumper is hooked up to its own hydrant and can quickly take up its position at the fire ground. If the fire continues to gain headway after the initial attack, other pumpers can supply water to the pump, which is already in position.

The forward lay permits the use of 1½-, 1¾-, or 2½-inch attack lines or a master stream if the hydrant is adequate and the proper diameter supply line(s) has been laid to supply the pumper.

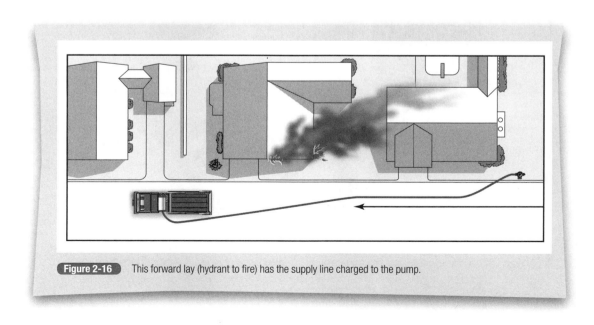

Figure 2-16   This forward lay (hydrant to fire) has the supply line charged to the pump.

Key Points

The most desirable hose lay is the forward lay from a hydrant to the fire, with the immediate charging of the supply line to the pumper, which enables the engine company to function independently with an uninterrupted supply of water from the beginning of the operation.

The following are advantages of a forward lay:
- The engine company is self-sufficient in terms of water supply, unless the fire is extremely large and the pumping capacity of additional engines is required.
- The company is free to take up any position in the front, rear, or side of the burning structure and begin operations with an uninterrupted water supply.
- Another responding engine company may drop off its crew and use hose lines taken off the forward engine company. If

needed, this later arriving engine could then lay an additional supply line to another hydrant(s), supplementing the water supply on the fire ground.
- A large hose line (2½-inch line) or a preconnected master stream appliance may be used almost immediately after arrival at the fire, with an uninterrupted water supply.

The following are disadvantages of a forward lay:
- There is a slight delay while the pumper stops and drops off a fire fighter to dress the hydrant with the supply line.
- There is a temporary loss of this fire fighter, who must remain at the hydrant until the supply line is charged.
- Supply lines being laid may hinder access of ladder company or aerial apparatus.

Forward Lay Using an Uncharged Supply Line

In this approach, a supply line is laid from the hydrant to the fire, but the line is left uncharged at the hydrant for another company to charge or hook up to **Figure 2-17**. Now the entire crew of

First pumper drops line at hydrant.

Second pumper supplies first pumper.

Figure 2-17 This forward lay (hydrant to fire) has the supply line uncharged until the arrival of the second company.

the first pumper can proceed to the fire; the fire fighter who wraps the supply line around the hydrant boards the pumper immediately and proceeds with the rest of the crew to the fire ground. Another company will now be responsible for going to the hydrant, hooking up the supply hose and charging the line. Two companies are now required to provide one pumper with a water supply to the fire ground. When the second pumper arrives, it will stop at the hydrant and charge the supply line before it is assigned another task.

This operation is popular when the second or later-arriving pumpers can be expected to arrive soon after the first. Standard operating guidelines and radio communications between engine companies can help ensure that the second engine company is aware of the situation and will respond to the location and make certain the supply hose line is charged. If pumpers do not arrive in fairly rapid succession, the first pumper will have to work off its water tank. The problem may become critical if the pumper runs out of water. This operation is also useful when narrow driveways or limited access may hamper operations at the fire building.

If the first-arriving officer notes that the fire is serious or if smoke or fire is showing, he or she must ensure that the company has laid and charged its own supply line unless extenuating circumstances dictate otherwise.

The following are advantages of laying a supply line and leaving it at the hydrant to be charged by a second pumper:

- After laying the supply line at the hydrant, the entire crew can proceed to the fire scene to fight the fire.
- The second-arriving pumper should arrive shortly thereafter and establish a water supply.

The following are disadvantages of laying a supply line and leaving it at the hydrant to be charged by a second pumper:

- Two engine companies are required to provide one engine company with an uninterrupted water supply.
- Another company must respond from the same direction for this task to be completed in a timely manner.
- At least two engine companies must arrive at the fire ground in quick succession if this procedure is to be efficient.
- If radio communications are not monitored or standard operating guidelines are not followed, there is a good chance that this evolution will not be completed in a timely manner.

Key Points

The forward lay using an uncharged supply line is popular when the second or later-arriving pumpers are expected to arrive soon after the first, which requires successful radio communications between engine companies.

Direct-to-Fire—No Line-Laid Approach

In this approach, the first pumper to arrive at the fire ground lays no supply line. It proceeds directly to the fire and begins working off the water in its tank. This can be a safe and effective procedure only when at least two companies are sure of arriving close together or, better still, see each other arriving. The second pumper lays an adequate supply line from the first pumper to the hydrant and charges the line **Figure 2-18**. Another alternative would have the second pumper lay a supply line from a hydrant to the first pumper for a quick water supply. If needed, the second pump could then obtain its own water supply.

The direct-to-fire approach is quite effective in areas that do not have hydrant systems. One unit goes to work on the fire immediately, whereas the second engine lays a supply line from the operating pumper to the nearest water supply such as a pond, lake, stream, or pool. At this location, the second pumper sets up to draft. This allows some firefighting, and possibly rescue operations, to take place during the time it takes to lay the supply line and get the hard suctions hooked up for drafting. Other pumpers coming in can give their tank water to the first pumper or use it in their own hose lines, as the situation demands. If the second pumper cannot reach the water supply due to distance, other engine companies can set up a

Figure 2-18   In the direct-to-fire approach, the first pumper works from the water supply in its water tank. When the second engine company arrives, it lays one or two supply lines from the first pumper to a hydrant and pumps to the first company.

relay pumping operation from the water supply to the pumper at the fire.

The following are advantages of the direct-to-fire–no-line-laid approach:

- The first-arriving pumper and its entire crew go directly to the fire to begin firefighting operations.
- The crew of the second pumper is dropped off at the fire; they can use attack lines off the first pumper on the fire once a water supply has been established.
- This approach is effective when drafting is necessary to obtain an uninterrupted supply of water.

The following are disadvantages of the direct-to-fire–no-line-laid approach:

- The operation will not work unless companies arrive close together, communicate with each other, and follow standard operating guidelines.
- Two engine companies are required to provide one engine company with an uninterrupted water supply.
- There is little or no margin of safety for attack crews.

Reverse Lay Using a Charged Line

In the reverse lay, the pumper lays firefighting hose lines or an LDH supply line from the fire to the hydrant. The pumper is then hooked up to the hydrant and when ready, pumping operations can begin. **Figure 2-19** shows the actions of one engine company using the reverse lay.

This type of operation has one advantage not mentioned in the discussions of other hose lays: The pumper is moved away from the fire building. It may be important to position aerial ladder trucks or elevating platforms near the building to perform rescue and firefighting operations. There may be a situation of a severe fire condition or the possibility of structural collapse that would preclude fire apparatus being close to the building.

An engine company using a reverse lay may use a 2½-inch hose line with a playpipe and a 1⅛-inch smooth-bore tip. The tip should be a leader line type. This tip is equipped with a screw-down thread at the end. Attached to the end of the tip is 100 ft of 1½- or 1¾-inch attack line with an appropriate nozzle. In this manner, an engine company could begin firefighting operations with either size attack line.

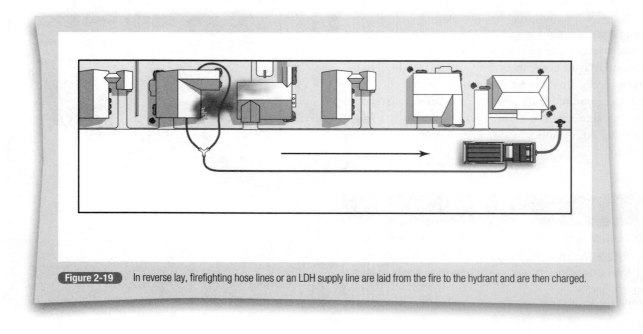

Figure 2-19   In reverse lay, firefighting hose lines or an LDH supply line are laid from the fire to the hydrant and are then charged.

A different approach may be to reverse lay using an LDH with a portable hydrant or manifold. In this mode of operation, the pumper goes to the hydrant while handlines are attached to the portable hydrant and operated on the fire.

The following are advantages of the reverse lay:

- The engine company is self-sufficient in terms of its own water supply.
- Later-arriving engine companies may initially hook up to a pumper operating from a hydrant and, if required, lay additional lines to the fire.
- Pumpers are not positioned close to the fire and thus do not block the approach and operation of aerial ladder trucks or elevating platforms.

The following are disadvantages of the reverse lay:

- This procedure does not get water onto the fire as fast as operations with preconnected hose lines.
- Equipment needed on the fire ground will have to be removed from the pumper before the pumper proceeds to the hydrant. Standard operating guidelines should dictate what fire fighters should remove from the pumper before it proceeds to the hydrant.

Wrap-Up

Chief Concepts

- The tools for fighting fire and providing emergency services include the fire apparatus and the equipment carried onboard the apparatus.
  - Pumpers and equipment should be purchased following recognized standards, federal requirements, and procurement policies set forth by the AHJ.
  - The purchaser should ascertain design, function, and performance requirements to assure that equipment received will function effectively, efficiently, and safely.
- The selection of fire apparatus and equipment should not be left to one or two individuals.
  - Sometimes fire chiefs or high-ranking officials lose touch with the everyday operation of the department.
  - Fire fighters should be queried as to their thoughts and ideas as to what type of tools and equipment may best serve the needs of the department.
  - A concerted effort should be made to purchase equipment that will serve the department for years to come.
- Standard operating guidelines should be established for all engine company tasks that need to be performed on the fire ground.
  - They include but are not limited to the following:
    - Water supply
    - Hose lays
    - Initial fire attack
    - Backup lines
    - Relay pumping
    - Tanker shuttle operations
    - The use of private fire protection systems
    - Property conservation
    - Overhaul
- A standard operating guideline is just that—a guideline for a company to use on a standard approach to a situation.
  - With standard operating guidelines in place, an engine company can commence operations that they, as well as other fire fighters, are familiar with.
  - Their actions help develop the overall emergency action plan.

- Because of the size and complexities of different fire departments and demographic considerations encountered, standard operating guidelines will differ from jurisdiction to jurisdiction reflecting the different needs of the department.
  - In the final analysis, for an engine company to be effective, it must have sufficient personnel that are well trained, properly equipped, and led by competent staff who operate using an incident management system.

Key Terms

Ball valve: Valves used on nozzles, gated wyes, and engine discharge gates. Made up of a ball with a hole in the middle of the ball.

Crosslays: Traverse hose beds.

Divided hose bed: A hose bed that is separated into two supply hose compartments running the length of the hose bed.

Double male and double female fittings: Used to connect two threaded connections of the same size and sex.

Hydrant assist valve (HAV): Also known as a four-way valve.

Master stream appliance: A large-capacity nozzle that can be supplied by two or more hose lines or fixed piping. It can flow in excess of 300 gallons per minute. Includes deck guns and portable ground monitors.

Prepiped master stream appliance: A master stream appliance that has a separate discharge pipe of adequate diameter that runs from the fire pump to the appliance.

Pumper fire apparatus: Fire apparatus with a permanently mounted fire pump of at least 750 gpm (3000 L/min) capacity, water tank, and hose body whose primary purpose is to combat structural and associated fires.

Supply hose: The hose used to deliver water from a source to a fire pump.

1. Standard equipment requirements for an engine company are _____.
   a. Much different in urban, suburban, and rural areas
   b. Somewhat different in urban and rural settings with hose and nozzle and ladder requirements varying greatly—other equipment is fairly similar
   c. Basically alike in urban, suburban, and rural areas
   d. Very similar for urban and suburban, but much different for rural

2. Time is the ally of the _____.
   a. Fire
   b. Fire department
   c. Fire fighter
   d. All of the above

3. NFPA _____ *Standard for Automotive Fire Apparatus* defines equipment requirements for new fire apparatus.
   a. 1001
   b. 1091
   c. 1591
   d. 1901

4. The absolute minimum size water tank for an engine is _____ gallons.
   a. 250
   b. 300
   c. 500
   d. 750

5. The 1¾-inch line preconnected hose line is suitable for
   a. Fires of limited size
   b. Fires of limited size, but multiple 1¾-inch lines are suitable for nearly all class A fires regardless of size
   c. Fires of all sizes, provided that the fire area is compartmented
   d. Fires requiring a rate of flow of less than 300 gpm

6. When the length of 1¾-inch hose is greater than _____ feet, a 2½-inch hose should be used to increase the length.
   a. 150
   b. 200
   c. 250
   d. The length of hose is not the critical variable, the maximum length is related to flow

7. A screw-down thread at the end of a solid-stream nozzle that allows a smaller line with 1½-inch couplings to be attached is known as a(n) _____ type tip.
   a. Extension
   b. Leader line

8. Comparing a 4-inch LDH to a 2½-inch hose: The 4-inch hose of equal length will flow approximately _____ more water than the 2½-inch hose.
   a. 1.5 times
   b. 2.5 times
   c. 3 times
   d. 3.5 times

9. An HAV
   a. Connects to the fire hydrant and permits the fire hydrant to be turned on, but will not flow freely until a back pressure is sensed in the supply line
   b. Attaches to the hydrant and allows another engine to augment the water supply without shutting down the hydrant
   c. Is a radio controlled valve operated by the pump operator, thus allowing the fire fighter at the hydrant to join the interior crew before the water is started to the supply line
   d. Is attached to the pump intake valve and permits the fire hydrant to be turned on, but will not open until the pump is engaged

10. The most desirable hose lay for supplying an engine with water is a
    a. Forward lay using a charged supply line
    b. Reverse lay
    c. Direct-to-fire–no-line-laid (using apparatus water tank as supply)
    d. All of the above

11. A *disadvantage* of a forward lay using an uncharged supply line is
    a. Two engine companies are required to supply one engine with an uninterrupted water supply.
    b. Another company must respond from the same direction to be effective.
    c. Two engine companies must arrive in quick succession to be effective.
    d. All of the above are disadvantages.

12. When faced with a severe fire condition or possible structural collapse, the _____ would generally be the best water supply.
    a. Forward lay using a charged supply line
    b. Forward lay using an uncharged supply line
    c. Reverse lay
    d. Direct-to-fire–no-line-laid approach (using apparatus water tank as supply)

Apparatus Positioning

Chapter 3

Learning Objectives

- Understand the concept of developing a preincident plan to determine resources and actions necessary to mitigate anticipated emergencies at a specific facility.

- Recognize the basic apparatus positioning assignments often defined in written standard operating guidelines.

- Consider that positioning of apparatus at the front, rear, and sides of the building, if possible, must be carried out to establish effective control of the fire.

- Understand that certain buildings may present positioning and coverage problems because of size, construction, location, and use.

Much depends on how engine companies are positioned at a fire ground. To a great extent, the positions taken up by pumpers as they arrive at a fire building are based on learning before the fire what problems may be encountered. This requires knowledge of the company's district, which can be acquired through preincident planning, inspection, driving throughout the area paying attention to such things as construction features, hydrant locations, and/or other sources of water, and access routes. The first section of this chapter discusses preincident planning, including the benefits of planning. The second section of this chapter deals with basic coverage responsibility. The next two sections deal with coverage in more detail and types of buildings that can pose coverage problems. Remember that standard operating procedures (SOPs) should mandate apparatus placement.

Preincident Planning

A **preincident plan** is a document developed by gathering general and detailed data used by responding personnel to determine the resources and actions necessary to mitigate anticipated emergencies at a specific facility.

NFPA 1620, *Recommended Practice for Pre-Incident Planning*, was developed with the primary purpose to aid in the development of a preincident plan to help responding personnel effectively manage emergencies with available resources and should not be confused with fire inspections, which monitor code compliance.

The preincident plan should be coordinated with an incident management system. The ultimate goal of preincident planning would be to know the problems involved in every potential fire building in a company's response area; however, if every building had to be preplanned, the process would be hopelessly time consuming.

Building Versus Structure

Fortunately, most of the buildings in an area are usually similar in many respects. For example, a neighborhood might consist mainly of two-family semidetached dwellings, with access to the rear of each building available through a driveway, and a hydrant within 200 ft on either side. Inspection of one such building is, in effect, preincident planning of them all. Although it might be helpful to examine each building in detail, the return would probably not be worth the effort.

Key Points

The ultimate goal of preincident planning is to know the problems involved in every potential fire building in a company's response area for each specific building (occupancy).

The buildings that must be examined individually and in detail are those that are unusual in some respect. Construction, unusual either in terms of dimensions or materials, warrants special consideration. The following factors should be evaluated when assessing the potential situations that could affect a facility during emergency conditions:

- Construction (access to buildings)
- Occupant characteristics (life safety considerations)
- Protection systems
- Capabilities of responding personnel
- Availability of mutual aid
- Water supply
- Exposure factors

In developing a schedule for preincident plans, strong consideration should be given to items such as the following:

- Potential life safety hazard
- Structural size and complexity
- Value
- Importance to the community
- Location
- Presence of chemicals
- Susceptibility to natural disasters

Such buildings present special positioning and coverage problems. Proper positioning of a pumper at a fire depends on several items:

- Physical characteristics of the *structure*
- *Location* of the structure
- Availability or lack of *water supply*
- Normal alarm *response* pattern with companies in quarters
- *Size and location* of the fire
- Need for aerial devices

This chapter later discusses some of the positioning and coverage problems involved with special types of buildings. The only way to determine whether such problems exist or whether positioning will be affected by special arrangements needed for, for example, a water supply, is to go to the site and take a look. If any special problem is suspected, an interior and/or exterior inspection are called for. This can be part of the regular fire safety inspection required by many municipalities; it can be part of preincident planning, but it should be done. No company responding to an alarm should receive any surprises at the fire ground.

Basic Coverage Responsibility

Apparatus positioning and coverage are often defined in a set of standard operating guidelines (SOGs), sometimes called SOPs. Department SOGs may differ from jurisdiction to jurisdiction. For example, a department may mandate that the first-in engine must lay a supply line while another allows the company to go directly to the building and operate on tank water. Generally, the company that is expected to arrive at the fire ground from its station first (the first due) is usually assigned to the front of the building. The first-due company's officer will assume command and assign apparatus according to the initial sizeup. Additional responding apparatus may be assigned a task or to an initial staging area. Companies should not be assigned a task unless there is a task for them to perform. Engine companies should not be allowed to "freelance" (i.e., independently decide what needs to be done or act outside of a superior's orders). The incident commander should assign apparatus to positions that will support the attack. A company should be assigned to a position that will allow it to perform the assigned task. The company officer should ensure that the apparatus is located in a safe position and, if possible, that other apparatus will have access to the area. It should be common practice for the first-arriving pumper to pull past the fire building. This accomplishes two items:

- Allows officers to see three sides of the building for sizeup
- Leaves room for the ladder truck in front of the building

Coverage assignments are usually based on the proximity of companies to the fire ground, but it is important to realize that they are only a guide for normal operations. No guideline should be used as a substitute for the judgment and initiative of the company officer. If the fire situation requires that the first-arriving company take a position at the rear of the burning structure, that is where it should be positioned; however, the officer should notify other responding companies, through their communications network, of any deviation from assigned positions. The second and additional responding pumpers will then be prepared to adapt to the situation. In this example, the second pumper would be advised that it should cover the front.

By the same token, an engine company out of its station ("on the air") and situated so that it will not arrive at the fire ground when expected should immediately communicate this fact to the dispatcher upon receiving the alarm. With this information,

coverage assignments can be modified; without it, part of the burning building might not be covered.

Whether a SOG is followed to the letter or is modified because of special circumstances, the end results are that coverage responsibility is distributed among responding pumpers and that each crew knows its own responsibility.

Coverage

This section discusses some general aspects of the complete coverage of fire buildings. The word "coverage" as used here means the assignment of companies to particular portions of the fire ground for sizeup and to accomplish any or all of the objectives of a firefighting operation. The first and most basic step in coverage is the positioning of the apparatus. The front of the fire building may indicate an entirely different situation than the rear or the sides **Figure 3-1**. Both front and rear (and the sides, when possible) must be covered quickly in order to establish effective control of the fire. For example

- First pumper: front of the building
- Second pumper: water supply/backup line
- Third pumper: rear of the building
- Fourth pumper: water supply for Rapid Incident Team (RIT)
- First ladder: front of building
- Second ladder: rear of building

The following examples provide a review of apparatus positioning at some of the most common types of buildings that a fire department may encounter. These are general recommendations. Depending on the type of building and any extenuating circumstances present, command must be able to develop an incident action plan to establish an overall operational approach.

Positioning—Front

The first-arriving company is normally assigned to cover the front of the building. In most situations, it will be obvious where the front of a building is located, but other times may be less obvious. Buildings may have more than one main entrance, or the entrance may be away from a street or parking area. When command designates the front of the building, all personnel must be notified. Building sides are designated A to D, with A (Alpha) usually assigned to the front. Letters continue clockwise around the building with the left side assigned to B (Bravo), the rear C (Charlie), and the right side D (Delta). This divides the building

Figure 3-1 Fire conditions vary from the front (*left*) and to the rear (*right*) of an involved structure.

into areas that are easy for command to manage. If resources are needed at the right side of the building in the rear, they would be directed to the C–D corner of the building. The position the first-arriving company takes may depend on the building type, as described in the following sections.

Narrow, Detached Building On arriving at a single-family detached dwelling, for instance, the first pumper should proceed just past the building. This permits the officer and crew to observe one side as they approach, then the front, and then the far side when they stop **Figure 3-2** . Only the rear will not have been observed, and that will be covered by the next-arriving engine company. Even if the next company to arrive is a ladder or rescue company, at least part of its crew should be assigned to check the rear. In Figure 3-2, the engine company, positioned in this manner, leaves room for a

later-arriving ladder company to get into position in front of the building. This approach and position are effective even when the building is relatively tall. The important point is that the first-arriving crew obtains a quick view of at least three sides of the structure.

Wide-Frontage Building Structures such as warehouses, garden apartments, large stores, and factories may be detached, but they usually have wide street frontage. In such cases, the first-arriving company can at least observe the approach side of the building and the front. The pumper should be positioned so that entrances to the building can be used in attacking the fire **Figure 3-3** . Later-arriving pumpers may be used to cover the rear, predicated on the strategic goals of command.

Key Points

The first-arriving company is normally assigned to cover the front of the building. The position it takes may depend on the building type.

Key Points

On arriving at a single-family detached dwelling, the first pumper should proceed just past the building, permitting the officer and crew to observe one side as they approach, then the front, and then the far side when they stop.

Figure 3-2 As the first pumper arrives at the front of a detached building, it should proceed just past the structure, observing one side, the front, and finally the other side during its approach.

Figure 3-3 In approaching a building with wide frontage, the pumper should be positioned at the entrance that is most accessible to the fire.

> ## Key Points
>
> When positioning in front of wide-frontage buildings, such as warehouses, garden apartment buildings, or large factories, pumpers should be positioned so that entrances to the building can be used in attacking the fire.

Attached Buildings   Fire in a building that is part of a continuing complex of attached structures presents a more acute problem. The sides of all but the end buildings are hidden, and the first-arriving company has a limited view of the situation. When the ladder company is approaching the fire building from the same direction, the engine company should position itself slightly past the front of the fire building. This will keep the front clear for ladder company operations. When the ladder company is approaching from the opposite direction, the engine company should stop short of the front of the building **Figure 3-4**.

Companies at the front of an attached building usually will have easy access to its interior. They will be able to enter quickly and make use of interior halls and stairways in multistoried

> ## Key Points
>
> In a complex of attached structures with the ladder company approaching a fire from the same direction as the pumper, the engine company should position itself slightly past the front of the fire building, keeping the front clear for ladder company operations. When the ladder company is approaching from the opposite direction, the engine company should stop short of the front of the building.

buildings. They can attack fires in these areas readily and get into ground floor areas easily. In standpiped buildings, the intakes are usually near the front entrance. Moreover, usually it is easier to ladder a building at the front.

Companies positioned at the front of the building can advance hose lines to attack the fire, get hose lines over the fire, and stretch backup lines as required by the type of structure involved, the fire situation, and the number of engine companies and personnel at the fire ground.

Positioning—Rear

As just described, covering the front of a building is usually no problem; the first-arriving pumper is positioned on the street that the building faces. Covering the rear may not be so easy, however, as positioning is affected by the layout of the block or area in which the fire building is located. Make sure that the engine lays a line going to the rear of the building. In most instances, a fire hydrant may be in the rear.

Detached Buildings   Companies assigned to rear coverage can and should use alleys, service roads, or driveways to get into position quickly. Engine companies should be positioned to allow room for ladder companies to get to the rear of the building. This may mean moving beyond the rear of the building or stopping short of it, depending on the direction from which the ladder truck would normally approach.

> ## Key Points
>
> Companies assigned to rear coverage can and should use alleys, service roads, or driveways to get into position quickly.

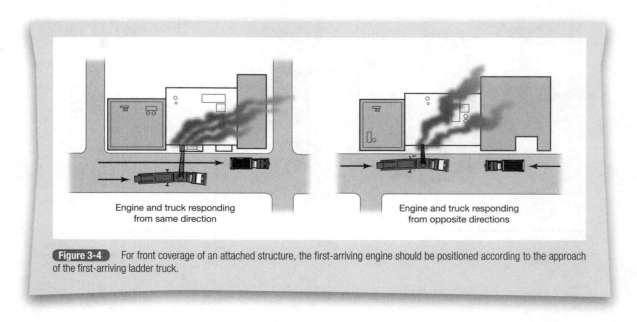

Engine and truck responding from same direction

Engine and truck responding from opposite directions

Figure 3-4   For front coverage of an attached structure, the first-arriving engine should be positioned according to the approach of the first-arriving ladder truck.

When one engine company has allowed space for ladder company movement, others should take care not to block this space. An engine should not be positioned in an alley in such a way as to prevent a ladder truck or personnel and equipment from getting to the rear of the building. Care must be taken not to block narrow alleys or other restricted passages that are the only available access routes.

In some residential, light commercial, or industrial areas with detached buildings, rear accessways may not be wide enough for a pumper. This means that the company or companies assigned to cover the rear must position their apparatus on the street and carry their hose lines and equipment to the rear **Figure 3-5** .

In some areas, driveways or parking lots permit positioning of the pumper close to the rear alongside the fire building **Figure 3-6** . This position can be taken if the driveway is not too close to the building and if fire and wind conditions allow such action.

Attached Buildings   Rear coverage is doubly important in attached buildings because the backs of buildings on adjacent parallel streets are usually fairly close to each other. Especially in apartment buildings, the rear windows, porches, and balconies must be checked for victims in need of rescue.

Even though all the buildings on a street are attached, there may be an alleyway behind them. In that case, the second-arriving

Key Points

If the pumper cannot be driven to the rear of an attached building, it must be positioned at the front or the end of the building row according to the location of the building on fire.

engine company can proceed up the alley, from the cross street, to the rear of the fire building **Figure 3-7** .

If the pumper cannot be driven to the rear of an attached building, it must be positioned at the front or the end of the building row according to the location of the building on fire. Fire fighters then can get to the rear through the fire building, an adjoining building, or perhaps a walkway. If the fire is located above the first floor, the crew may be able to pass to the rear through the fire building; however, this might require positioning the pumper where it would add to the congestion at the front of the fire building. In such a case, it would be better for the crew to use an adjoining building or one on the next parallel street that is back to back with the fire building **Figure 3-8** .

Engine companies at the rear might not find it as easy to get hose lines into position as they would at the front of the building, unless there are ample rear alleys. Actions necessary to get hose lines into position will vary with the situation.

Figure 3-5   When no rear access to a detached building is available, pumpers need to be positioned in front of the building so that hose lines and equipment can be carried to the rear.

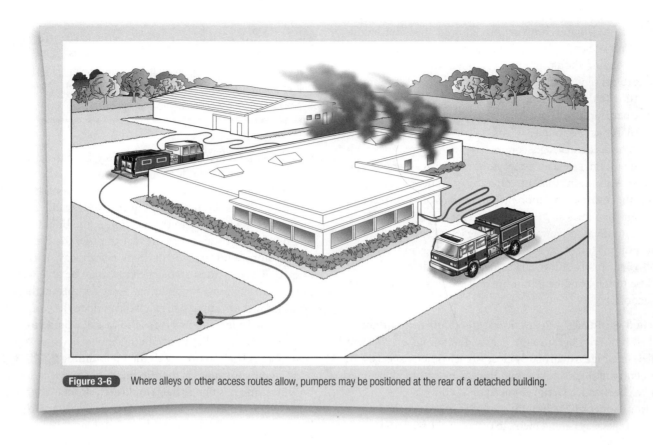

Figure 3-6 Where alleys or other access routes allow, pumpers may be positioned at the rear of a detached building.

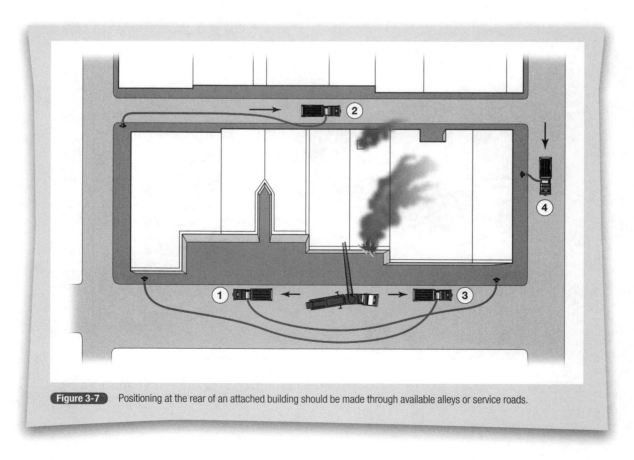

Figure 3-7 Positioning at the rear of an attached building should be made through available alleys or service roads.

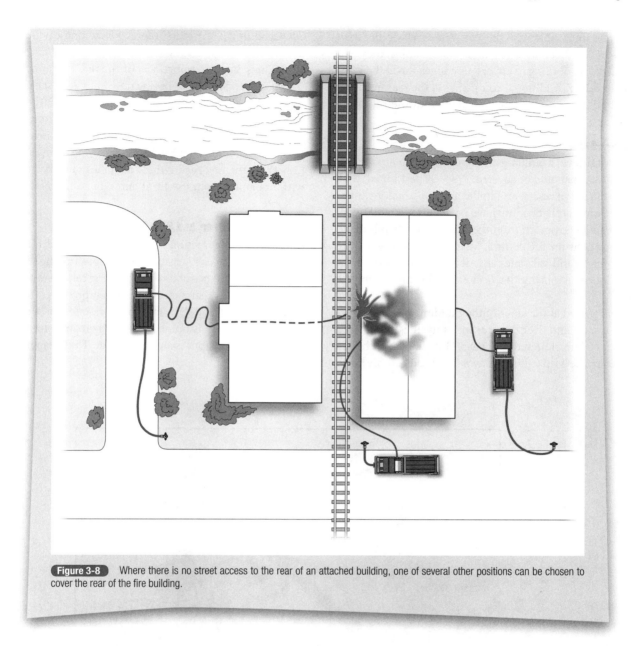

Figure 3-8   Where there is no street access to the rear of an attached building, one of several other positions can be chosen to cover the rear of the fire building.

Engine companies positioned at the rear of a building can advance their hose lines to attack the fire, get lines over the fire, bring in backup lines, access adjoining structures, gain quick access to basement entrances, and/or carry out any other required operation if there are enough personnel and engine companies at the scene. In buildings more than one story high, fire fighters can take advantage of rear interior stairs, rear exterior stairs to porches, and rear fire escapes, or they may use other methods of getting their hose lines into a building. These methods are discussed in Chapter 6.

In some cases, fire fighters of the second-arriving engine company entering a building from the rear might find it necessary to take their hose lines toward the front through a hallway and then up the same stairs used by the engine company

positioned at the front. This will depend on the depth of the building and its interior features. Even in this case, however, the engine company at the rear will have been able to check

Key Points

Engine companies positioned at the rear of a building can advance their hose lines to attack the fire, get hose lines over the fire, bring in backup lines, access adjoining structures, gain quick access to basement entrances, and/or carry out any other required operation if there are enough personnel and engine companies at the scene. Engine companies at the rear of the building will also support roof-top operation with protective hose lines.

the rescue and fire problems at the rear. Where a lay other than the reverse lay has been used for initial water supply, they will have avoided contributing to apparatus congestion at the front of the fire building. Additional companies at the rear may find other methods of getting their hose lines into the building, which will result in an efficient operation.

Positioning—Sides

If one or both sides of a building are exposed and companies are available, the sides should be covered. Properly performed, such operations can aid in rescue and initial attack operations. After interior stairways and corridors have been covered, additional hose lines can be advanced from the sides. This often leads to excellent positions for fire control. Side interior stairways, fire escapes, porches, and balconies are used to advance hose lines if such action does not hinder occupants escaping from the building.

Ladders placed at the sides of the building are used to get hose lines around and over the fire, and thus, these lines need not be advanced up cluttered stairways. Hose lines stretched in this manner can be used to knock down fire ahead of those being advanced down corridors, thereby lessening the punishment taken by the crew on those lines. If this tactic is to be used, companies must support a coordinated fire attack and be cognizant of operating opposing hose lines against each other. If possible, ladders (aerial and ground ladders) should be placed on all four sides of a building for access and egress from building.

In one-story structures, especially large, well-involved units, hose lines from the sides can be used to bring additional streams to bear on the fire. Again, these activities must be coordinated with the attack from the front and rear to promote efficiency and safety.

General Front, Rear, and Side Operations

Emphasis in this section has been on the need for complete coverage (insofar as is feasible) and the positioning of apparatus so that as much as possible of the building can be seen. Complete coverage assures that exposures will be protected by properly laid hose lines **Figure 3-9**. In terms of rescue, the benefit of this complete coverage is immediate. The benefit in terms of other firefighting operations is almost as great. These benefits include the following:

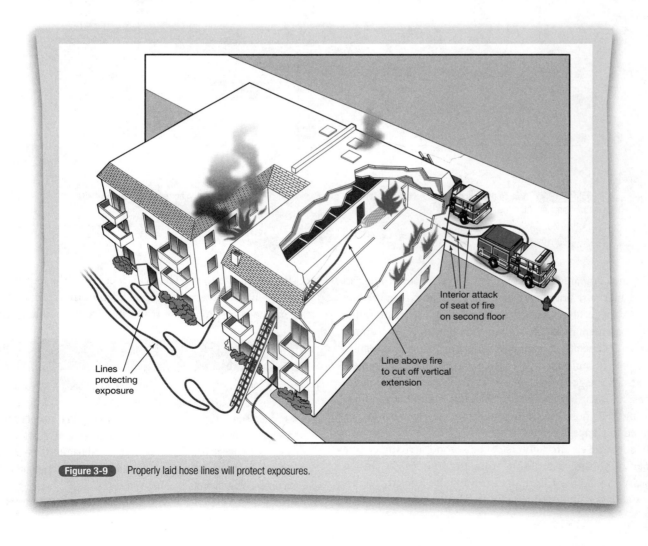

Interior attack
of seat of fire
on second floor

Line above fire
to cut off vertical
extension

Lines
protecting
exposure

Figure 3-9 Properly laid hose lines will protect exposures.

- The full extent of the fire will be known.
- There will be few, if any, surprises.
- Severe exposure problems will be discovered and dealt with.
- Companies surrounding the fire building can work in unison, assisting each other for maximum efficiency.

Problem Buildings

Some buildings might fit into one or more of the categories discussed in the last section, but they may present additional positioning and coverage problems because of their size, construction, location, or use.

Set-Back Buildings

A detached structure might be set far off the street, with a long drive leading to it and no room in which to maneuver apparatus (Figure 3-10). There may be other barriers or hindrances such as trees, a wall, or fence that block the proper positioning of apparatus. In such a case, the first engine company to arrive should approach as close to the building as fire and wind conditions permit. The second engine company should locate to the rear of the first engine, in a position that will not interfere with operation of the first engine. The crew of the second engine should then proceed on foot to provide rear or side coverage. This is best done by first moving to the far side of the building, away from the apparatus, and then moving on to the rear. The second crew thus has a chance to size up the far side if this could not be done from the road as they arrived.

Shopping Malls

Huge enclosed shopping malls are usually close to rectangular in shape. Most are designed to allow direct entrance into the central mall or court area through large doorways on either side of the structure. Although few doors are big enough for apparatus, hose lines can be carried through with little problem. This allows coverage of the front of the individual stores or offices in the mall, as most businesses front on the central court. The rear

of individual stores and offices is covered from the side of the structure in which they are located (Figure 3-11).

Malls are usually surrounded by parking lots or roads. Thus, there is ready access to the sides and entrances of the mall for front and rear coverage of a fire in a store or in the central court itself. When standpipes are not provided, unusually long attack lines will be necessary. Companies should be aware of such situations and should prepare and practice effective operations. Such preparation should include interior positioning of 1½-, 1¾-, and 2½-inch hose lines and master stream appliances. Preincident planning will certainly benefit fire fighters if a fire should occur in a mall.

Standard Shopping Centers

Standard shopping centers usually consist of attached blocks of buildings. Front and rear coverage here is as important as in the situations discussed earlier. Sometimes rear doors and windows permit ready access to the back of the building. In other cases, rear windows are limited in size, are barred, and/or are located high on the wall, and doors are made of steel and securely fastened from the inside; however, most windows can be put to use for fire attack or venting, and it might be possible to force the doors without losing much time. If conditions warrant, walls may be breached to gain entry, perform rescue, or attack the fire.

Mercantile Areas

Rear coverage is vital in mercantile areas in general. Many of these fires start in the work/storage sections behind the sales floors (i.e., in the rear of the store). Utilities (water, gas, and electricity) pass from store to store at the rear and pierce the walls. Fire can spread rapidly through these utility openings. Fire companies positioned at the rear may be the first to notice this fire spread and take action to stop it.

The main body of fire in the rear of the store, however, should be attacked primarily from the front or unburned side of the building. This tends to contain it in the rear and to protect undamaged display merchandise.

Garden Apartments

Front and rear coverage are most important in the non–fire-resistant structures commonly known as "garden apartments." Usually limited in height to three or four stories, these are actually frame buildings with a wood, brick, or masonry veneer. Often, the workmanship and materials are flimsy, and there is little to slow the spread of fire once it starts within the building.

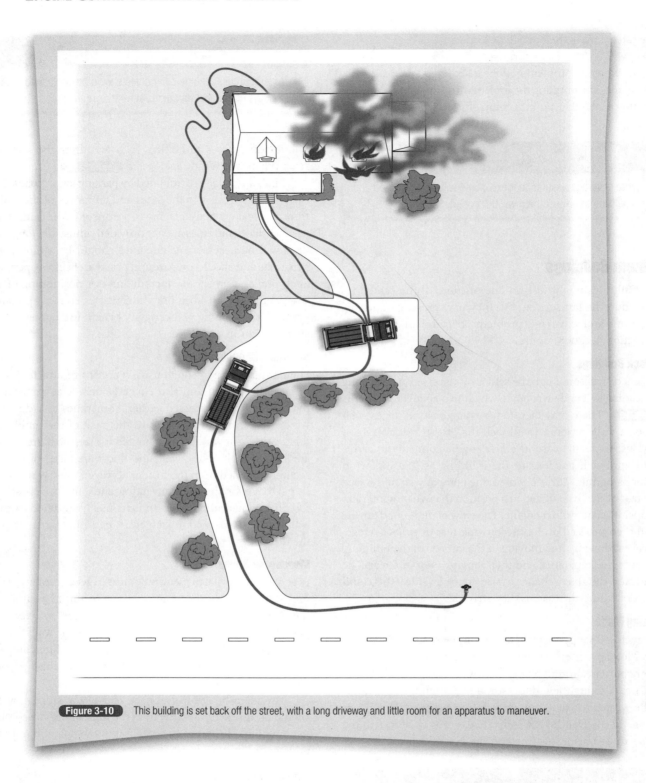

Figure 3-10   This building is set back off the street, with a long driveway and little room for an apparatus to maneuver.

Adding to the problem of poor construction is the manner in which these buildings are laid out. In the usual design, half of the apartments can be seen only from the front, and half only from the rear. Thus, the first-arriving company may not even notice a fire in a rear apartment **Figure 3-12**. If the interior design does not permit passage through the building from front to rear, occupants in trouble in rear apartments will not be seen or be reachable from the front. For this reason, it is important that command assigns other responding companies to the rear and that a sizeup is quickly made of this area.

Another problem is the varying height of some garden apartments. Building codes usually limit these buildings to three

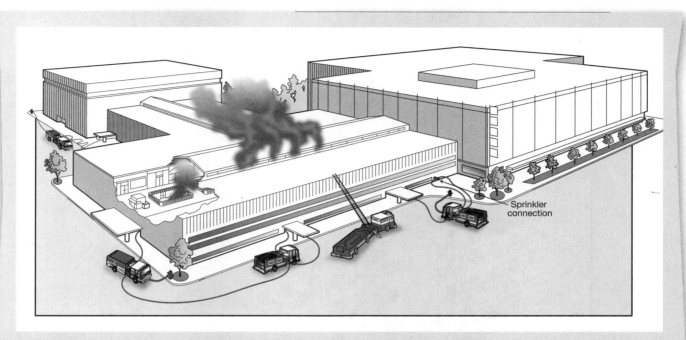

Figure 3-11　In covered malls, engine companies should be positioned to take advantage of the best entrances in relation to the fire. Also, protective systems should be located as soon as possible, in the event they need to be supplied.

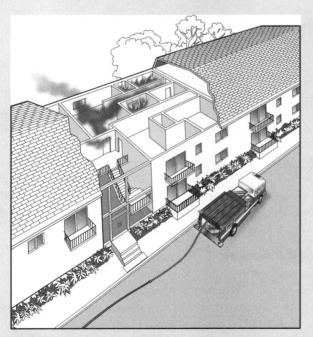

Figure 3-12　Because of the interior design of garden apartments, fire in a rear unit might not be readily detected by the first-arriving engine company.

Key Points

In the usual design, half of the garden apartments can be seen only from the front, and half only from the rear. Thus, the first-arriving company may not even notice a fire in a rear apartment. For this reason, it is important that command assigns other responding companies to the rear and that a sizeup is quickly made of this area.

and sometimes four stories; however, this is often measured from the position of the front entrance. If such a building is set into a steep grade, additional stories may be added at the rear. This means that fire fighters in front may be looking at a three- or four-story building, whereas crews in the rear must cope with a five- or six-story structure.

Occasionally, a garden apartment building that is extremely long is divided in two lengthwise, blocking access to the rear through the building. Also, the outside accessway to the rear may be too narrow for an engine. If the fire is in a rear apartment near either end of the building, there is no problem, but when it is located 200 or 300 ft or more toward the center, crews must have some way to get into action quickly. In some cases, lines can be stretched around to the back from pumpers positioned at

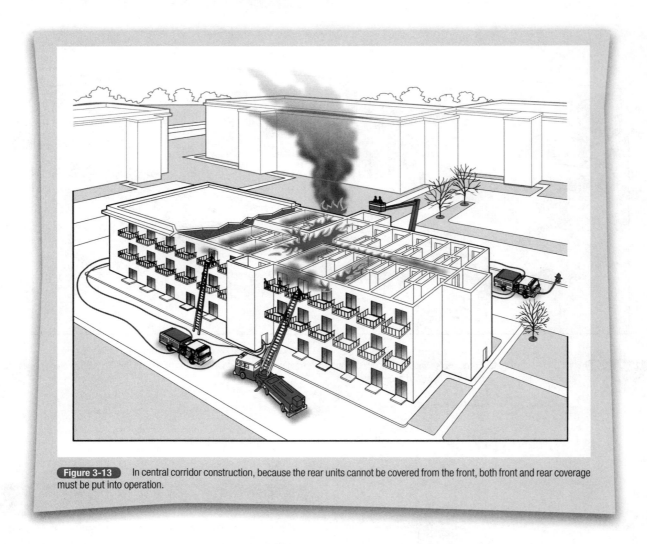

Figure 3-13 In central corridor construction, because the rear units cannot be covered from the front, both front and rear coverage must be put into operation.

the ends of the building. If extremely long stretches are required, it might be better to drop some fire fighters at the end of the building so that they can proceed on foot to the fire area. The engine is then located in the front, opposite the fire, and lines are passed over the roof by quickly laddering the front side. Many fire departments have had success with such operations, but only after planning and onsite training.

Central Corridor Construction

Other apartment houses and most motels, hotels, and office buildings are laid out differently, but with designs presenting similar problems. In these buildings, a central corridor on

> ### Key Points
>
> In a central corridor construction, because the rear units cannot be covered from the front, both front and rear coverage must be put into operation.

each floor divides the floor in two. Again, the result is that half the rooms can be seen only from the front and half only from the rear. Because the rear units cannot be covered from the front, both front and rear coverage must be put into operation **Figure 3-13**. This means that the rear of the building must be checked as quickly as possible. Depending on the situation, the sides of the building may also need to be covered. The first-arriving company is usually positioned close to the front entrance, because that is the major access point.

High-Rise Buildings

The first-arriving company should be positioned so that crew members have ready access to the front entrance and to standpipe inlets and outlets. Other responding engine companies are positioned by command at the rear and sides, if possible, with consideration to the width and length of the building and the locations of standpipe intakes, rather than by its height. Operations in high-rise structures are discussed in detail in Chapter 10.

Chief Concepts

- Development of a SOG for positioning of apparatus and for coverage of a fire ground begins with a good knowledge of the district or area of responsibility.
- The SOG itself should clearly place responsibility for coverage of a particular part of a building with each first-alarm company.
- Command officers must be capable of modifying the SOGs when the situation warrants it.
- Ordinarily, the first-arriving engine company covers the front of the fire building; the second company covers the rear, and other responding companies should be directed by command, ensuring that all sides of the building are covered.
- Availability of immediate access to the rear of a structure depends on whether the structure is detached, attached, wide, or narrow; near the street or set back; or served by rear alleys, driveways, or walkways.
- In addition, access to the rear and sides may be impeded by peculiarities of construction and/or use. These peculiarities tend to make full coverage of a fire building all the more imperative.
- Some buildings, such as garden apartments and high rises, present additional positioning and coverage problems because of their size, construction, location, or use. These buildings may require specific preincident planning.

Key Term

Preincident planning: The process used to gather information to develop a preincident plan.

1. It is recommended that _____ be preincident planned.
   a. All buildings
   b. All buildings except one- and two-family detached dwellings
   c. Only large buildings with an extreme life hazard or a potential for a large loss fire
   d. Buildings that are unusual in some respect

2. Standard operating guidelines
   a. Must be followed exactly as written
   b. Should be followed in all cases, but minor variations are allowed
   c. Are only examples of good operations; company officers should do whatever they think is best
   d. No guideline should be used as a substitute for the judgment and initiative of the company officer.

3. The first-arriving company is normally assigned to
   a. The side of the building where fire or products of combustion are visible
   b. The front of the building
   c. The rear of the building
   d. The sides of the building

4. Using the alphabetic system in designating sides of the building where the front of the building is designated as side A, the left side of the building would be side _____.
   a. B
   b. C
   c. D
   d. All sides of the building are designated when a formal command post is established, with the command post being side A, and all other sides specifically named by the incident commander.

5. The first pumper to arrive at the scene of a structure fire in a narrow (detached) building should position the apparatus
   a. On the near side of the incident short of the main entrance in the direction of their travel
   b. On the far side of the building, past the front of the building in the direction of their travel
   c. So that entrances to the building can be used to attack the fire
   d. Either slightly passed the building or short of the building depending on the direction of travel of the ladder company

6. The first pumper to arrive at the scene of a structure fire in a wide-frontage (detached) building should position the apparatus
   a. On the near side of the incident short of the main entrance in the direction of their travel
   b. On the far side of the building, past the front of the building in the direction of their travel
   c. So that entrances to the building can be used to attack the fire
   d. Either slightly passed the building or short of the building depending on the direction of travel of the ladder company

7. The first pumper to arrive at the scene of a structure fire in a row of attached buildings should position the apparatus
   a. On the near side of the incident short of the main entrance in the direction of their travel
   b. On the far side of the building, past the front of the building in the direction of their travel
   c. So that entrances to the building can be used to attack the fire
   d. Either slightly passed the building or short of the building depending on the direction of travel of the ladder company

8. When confronted with a row of attached buildings, the second-arriving engine company should position
   a. At the rear of the row of buildings using alleys or service roads
   b. At the front of the building behind the first-arriving engine company
   c. At the side of the building at the far end in their direction of travel
   d. At the side of the building at the near end in their direction of travel

9. In general, when there is access to all sides of a building, engine companies should be placed
   a. At the front, then sides, and finally the rear of the building
   b. At the front, then rear, and then sides of the building
   c. At the rear, then front, and finally at the sides
   d. There is no priority order for apparatus placement by side of the building.

10. When confronted with a detached structure that is set back from the street with no rear or side access for apparatus, the second-arriving engine company should position
   a. At the rear of the row of buildings using alleys or service roads
   b. At the front of the building behind the first-arriving engine company
   c. At the side of the building at the far end in their direction of travel
   d. At the side of the building at the near end in their direction of travel

11. In the case of a shopping mall, there is usually
   a. Good access to the front via the main entry doors, but very limited access to the rear of the mall
   b. Good access to the front via main entry doors and good access to the rear via loading docks
   c. Good access all around the building
   d. Poor access to front and rear due to setbacks and landscaping

12. Garden apartments are normally
   a. Constructed with back-to-back apartments
   b. Of limited height, usually three of four stories
   c. Of frame construction
   d. All of the above are generally true for garden apartments.

13. A fire in the rear of a(n) _____ may not be visible at the front due to the interior design.
   a. Office building
   b. Motel
   c. Garden apartment
   d. All of the above

Chapter 4

Learning Objectives

- Comprehend the chronology of rescue, which begins well before the alarm is received, is continued from dispatch to arrival, and is carried on at the fire scene from sizeup to the completion of the secondary search.

- Examine factors that affect search and rescue procedures in several types and sizes of occupancies, including fire-resistant construction.

- Recognize that a thorough, planned primary search should be conducted at every fire and that all fire fighters, no matter to what company they are assigned, should be able to conduct this search if needed.

Introduction

Operating at emergency incidents poses an inherent risk of injury or even death. All members operating at incidents must function in a safe manner. Each member must maintain a level of awareness for their own well-being, as well as to minimize the risk to others. Toward that goal, all members are expected to operate under a risk management profile. Every fire department should incorporate a standard operating guideline (SOG) that describes the operating policy regarding risk assessment and safety management at all emergency incidents. The SOG should state the following:

- Fire fighters will risk their lives a lot in a calculated manner to rescue savable lives.
- Fire fighters will risk their lives a little in a calculated manner to rescue savable property.
- Fire fighters will not risk their lives at all for lives or property that is already lost.

To ensure that this guideline is in place at emergency incidents, the following must take place. Command must be established. Fire fighters need to be in full personal protective equipment. An accountability system must be established. Safety procedures must be in place and adhered to, and there should be a continuous risk assessment conducted throughout the incident.

Every fire fighter, from the newest recruit to the seasoned veteran, knows that the rescue of endangered persons is the primary objective at the scene of a fire; however, not every fire fighter realizes how broad the scope of rescue operations actually can be.

Carrying a fire victim to safety is rescue work in the purest sense. Placing a ladder for use by entrapped people and assisting or directing people to leave a fire building are also rescue actions, as is searching a building for victims. Each of these is a rescue operation because it immediately and obviously reduces danger to human life.

Proper placement of the first hose lines can keep a fire away from people in the building. Rapid ventilation removes accumulations of smoke and gas and prevents their further buildup. Both operations reduce the danger to people inside and extend the time they have to get out of the building. In a very real sense, these, too, are rescue operations.

Residential fires, especially in single- and two-family dwellings, very often require rescue operations. More fire fighters are injured in single-family dwellings more than any other type of occupancy. National statistics show that injuries and deaths in residential fires far outnumber those in other occupancies, such as hospitals, schools, and nursing homes. This chapter discusses rescue operations, beginning with preincident planning. Then the primary search and fire attack are each considered in greater detail.

The Chronology of Rescue

Before the Alarm

Preparation for rescue begins well before the alarm is received. It begins with a thorough knowledge of the company's area of responsibility, including the occupancies, the hazards, and the potential rescue problems. This knowledge is based on building inspections, preincident planning of particular buildings, and being aware of changes that occur in the company's district. The objective is to know beforehand the approximate type and extent of rescue operations that could be involved in any fire.

Receipt of the Alarm

The alarm itself can be an indication of the potential rescue situation. The type of occupancy and the time of day, for example, are clues to the need for, and possible extent of, rescue work.

Fires in residential properties, from single-family homes to large apartment houses, always include the possibility of a rescue situation. The possibility is much greater and the problem perhaps more acute at night and in the early morning hours when most people are at home and asleep.

Buildings such as offices, schools, and large stores present a reverse situation. These buildings are normally empty during the night but have many people occupying them in the daytime. Here, it is the daytime fire that will almost surely require rescue operations.

Even the information given with a verbal dispatch can be important. Such phrases as "across from," "at the rear of," "next door to," and "near the intersection of" may indicate that the fire alarm was not turned in by an occupant of the involved building, but rather by someone outside the building who only saw smoke or fire and does not know what the situation is inside. These phrases frequently indicate trouble.

Key Points

Preparation for rescue begins well before the alarm is received. It begins with a thorough knowledge of the company's area of responsibility, including the occupancies, the hazards, and the potential rescue problems.

Key Points

The alarm itself can be an indication of the potential rescue situation. The type of occupancy and the time of day, for example, are clues to the need for, and possible extent of, rescue work.

The important point is not that these clues exist but that they are recognized. In particular, when the engine company arrives at the fire ground, the clues can be used as part of the information on which the initial sizeup is based.

At the Fire Scene

Even before the pumper has stopped, the officer should have begun a careful sizeup of the fire scene. The following questions can help to determine the extent of required rescue operations:

- Is the fire structure a closed-up house with heavy smoke showing?
- Are cars parked in the driveway, indicating that an entire family might be inside?
- Are people at the windows, or are other occupants calling for help?
- Are there indications that other victims might be inside, unable to get out of the building?

Bystanders can give unreliable information.

Figure 4-1 and **Figure 4-2** illustrate what a fire fighter should look for to determine whether people are still inside the fire building.

In addition, the extent of the fire and the occupancy, size, age, and apparent population of the building are important in ascertaining what rescue operations are needed. Some of this information would have been gathered during preincident planning or building inspections.

Other information useful in sizing up the situation might be volunteered by or solicited from neighbors and occupants who have escaped the fire building. Of special urgency is any report that there are still people inside. On the other hand, unofficial reports that "everyone has escaped" should not delay or halt the primary search, especially in multifamily residences. Because there is a limit to the time available for sizeup, it must he accomplished quickly and efficiently if it is to be of any use. Again, the more information available before the alarm, the more accurate the initial sizeup will be.

Initial sizeup will indicate where fire fighters should begin their search and rescue and fire attack operations. These

Key Points

Sizeup must be accomplished quickly and efficiently if it is to be of any use. The more information available before the alarm, the more accurate the initial sizeup will be.

Figure 4-1 Cars parked near the house, a bicycle and wagon nearby, and a bystander pointing toward the house indicate that there may be people in the building.

Figure 4-2 Victims at windows often indicate that others may be trapped inside the fire building.

operations, coordinated by the incident commander, should begin immediately.

Immediate rescue, without a coordinated fire attack, should be attempted only in extreme cases. Occupants attempting to jump from upper floors would constitute an extreme case of immediate rescue. In such a situation, the engine company would go directly to the fire building and affect the rescue. Additional responding pumpers would be responsible for establishing a water supply and stretching hose lines.

If victims are on upper floors at windows, fire fighters may assist victims by raising ladders and/or attempting to talk to the victims to calm them until they can be brought down **Figure 4-3** . Battery-powered megaphones are useful in such situations. Many times the victims that you see may not be the ones in dire distress.

Engine companies can aid rescue by placing protective streams between the victims and the fire, thus driving the smoke and fire away from the occupants. Any positive action on the part of arriving fire fighters will usually have a calming effect on those trapped in the building.

Two-In/Two-Out Rule

Fire fighters must be aware of the two-in/two-out rule as directed by OSHA and NFPA 1500. During the initial stages of an incident, fire fighters working inside the hazard area must work with a crew of at least two members. They must be backed up by at least two members outside the hazard area who are fully protected and able to go to the aid of the two members inside if the need arises.

Key Points

Initial sizeup will indicate where fire fighters should begin their search and rescue and fire attack operations. These operations, coordinated by the incident commander, should begin immediately.

Key Points

Two-in/two-out rule: During the initial stages of an incident, fire fighters working inside the hazard area must work with a crew of at least two members who must be backed up by at least two members outside the hazard area who are fully protected and able to go to the aid of the two members inside if the need arises.

Figure 4-3   Victims attempting to jump from upper floors indicate an immediate rescue situation.

Rapid Intervention Team

A **rapid intervention team (RIT)** must be provided from the initial stages of an incident to its conclusion. The objective of the RIT is to have a fully equipped rescue team on site, in a ready state, to immediately react and respond to rescue injured or trapped fire fighters and civilians. If possible, the RIT should be a crew of four fire fighters, one of whom must be an officer.

This process begins with the first-arriving companies using the two-in/two-out protocol. A minimum of two fire fighters should be assigned, but usually a company consisting of three or more fire fighters is assigned to this responsibility. If the incident becomes complex, additional companies should be assigned. RIT members should be in full Personal Protective Equipment (PPE) and have all necessary tools and equipment needed, including search rope and rescue equipment, lights, and EMS equipment. Other rescue equipment that may be considered are ground ladders, power tools, ventilation equipment, and, if needed, protective hose lines.

RIT members should stage on scene in a location to maximize their options and await instructions from command. During large operations, the RIT will normally be assigned near the command post. Members should closely monitor the assigned fire-ground radio channel at all times. RIT then should access the preplan off of the first engine and bring to the command post.

A SOG must be created to ensure that an RIT is established when members are engaged in active firefighting activities, or other incidents where fire department members are subject to hazards that would be immediately dangerous to life and/or health in the event of an equipment failure, a sudden change in fire-ground conditions, or a mishap.

Water is as essential to the primary search and rescue operations as it is to extinguishment. Water from attack lines is used to

- Separate the fire from the occupants closest to it by placing hose lines in strategic locations
- Control interior stairways and corridors for evacuating occupants and advancing fire fighters **Figure 4-4**

Key Points

Water is as essential to the primary search and rescue operations as it is to extinguishment.

Figure 4-4   Control of interior stairways and hallways must be established.

• Protect fire fighters performing the primary search, ventilation, and extinguishment around, above, and below the fire

Engine companies at the scene must therefore quickly set up to provide the necessary hose lines and provide a continuous supply of water to sustain the attack. The 500 gallons or more of water carried in the average-size water tank may not be sufficient for rescue operations. Supply lines should be laid by first arriving pumpers at the fire scene to assure a continuous flow.

If no hydrant system is available, water from the water tanks of later-arriving pumpers or mobile water supply apparatus should be fed to the first-arriving pumper **Figure 4-5**. Should later-arriving pumpers run their own attack lines, the first pumper, whose crew might be engaged in rescue operations,

Key Points

It is important to keep the first-arriving pumper supplied with water so that its crew can maintain their position and continue rescue and fire attack operations.

Figure 4-5   If no hydrants are available, the initial attack should be made by the first-arriving pumper (*at right*) with hose lines supplied by the tank water. The second-arriving pumper (*at left*) should supply the first pumper from its water tank.

will probably run out of water, and their operation and safety will deteriorate. It is important to keep the first-arriving pumper supplied with water so its crew can maintain their position and continue rescue and fire attack operations.

Stream Placement

Streams should be put into service as soon as possible to attack the main body of the fire. The primary function of the engine company in a rescue situation is to support the primary search, contain the fire, and keep it from jeopardizing anyone within the fire building. For first-arriving companies, this may mean keeping the fire from the occupants and, if necessary, allowing it to spread in a direction that would not endanger lives. When enough personnel become available, the attack should shift to the main body of fire; until then, the building is secondary. Human life must be protected even if it is necessary to sacrifice the building.

Streams must be placed carefully at this point. They must keep steam, smoke, and gases, as well as the fire, away from occupants, and they must not hinder fire fighters who may already have begun the primary search or are performing other firefighting tasks.

Figure 4-6 illustrates hose line positioning that establishes an effective stream between the fire and trapped victims. This positioning is the best rescue procedure for engine companies.

Primary Search

The primary search is the first search. If there is any indication at all that victims might be trapped or overcome within the fire building, search operations should begin as soon after arrival as

Key Points

The primary function of the engine company in a rescue situation is to support the primary search, contain the fire, and keep it from jeopardizing anyone within the fire building.

Key Points

Streams must be placed carefully to keep steam, smoke, and gases, as well as the fire, away from occupants. They must not hinder fire fighters who may already have begun the primary search or are performing other firefighting tasks.

Figure 4-6   The correct hose line position ensures that an effective stream is established.

possible. This will usually be at the time the first hose lines are being stretched and positioned.

The search for victims is normally the function of ladder or rescue companies. The engine company advances its hose lines and places streams into operation to support the primary search; however, if a ladder or rescue company is not available or has not yet arrived, engine company personnel must begin the search. (Search patterns are discussed in detail in the last section of this chapter.) No matter who performs the primary search, it is extremely important that every fire fighter at the scene be aware that a search is in progress. All activity should be directed toward supplementing the efforts of fire fighters engaged in the search and toward providing protection for them and for any victims they might find.

Secondary Search

A primary search is a quick but thorough search of an area to determine whether victims are still in the building. A secondary search will need to be conducted after the primary search to determine that no one was overlooked during the primary search. Both searches should be conducted in a systematic pattern so that no area is overlooked.

Ventilation

The building should be ventilated as soon as possible to allow smoke, heat, and gases to move away from occupants who might be trapped inside. On the fire ground, ventilation must be performed as part of the overall coordinated fire attack operation. When the fire is free burning, ventilation should begin at the same time as the initial attack. If the fire is smoldering, the building must be vacated properly before the building is entered. Ventilation must be coordinated with fire attack, especially when using positive pressure ventilation.

Fire Attack for Rescue

The construction of the fire building, its size, its occupancy, and its layout affect rescue operations because they affect the number of people inside and the paths that can be used to reach them. This

section discusses fire attack in conjunction with search and rescue in several sizes of occupancies and in fire-resistant construction. In any type of structure, the main thrust of operations is determined by the location and severity of the fire and the location of the occupants most endangered by it. The goal is to get the situation under control as quickly as possible.

Single-Family Dwellings

In a typical two-story, single-family dwelling with two or three rooms on fire on the first floor, the occupants in most danger are those close to the fire on the first floor and those directly over the fire on the second floor. The former will be affected by radiant heat and the latter by the convected smoke, hot air, and gases.

The main body of fire should be attacked immediately with the proper size of attack line. At the same time, fire fighters should be sent to the area over the fire to begin ventilating and searching for possible victims. No time should be lost in getting fire fighters to the upper floor. If an attack line is immediately available, it should be taken upstairs. If not, the primary search should begin without it.

Figure 4-7 illustrates a two-story, single-family dwelling with a fire on the first floor. The fire should be attacked and cut off from traveling to the upper floor, and a search of the upper floor should also be conducted, beginning with the rooms directly above the fire.

If fire is discovered in any upstairs room, the door should be shut and the room isolated until a hose line can be advanced. Doors and windows of other rooms should be opened to provide ventilation, providing that this will not spread the fire. Opened windows will dissipate some of the heat and smoke and thus allow more efficient search and fire attack operations. As the attack on the fire progresses on the first floor, the area around the fire should be searched thoroughly.

Multiple-Family Residences

For any large, occupied residential building, the location of the fire and the smoke above it should be carefully noted during sizeup. The

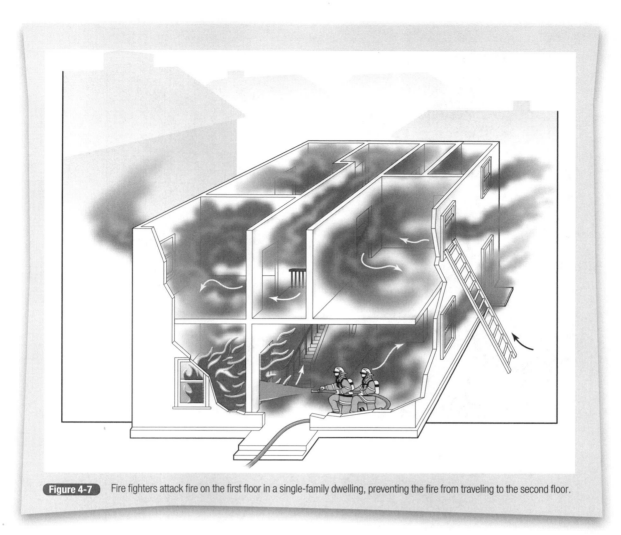

Figure 4-7 Fire fighters attack fire on the first floor in a single-family dwelling, preventing the fire from traveling to the second floor.

smoke indicates the area into which the fire will most likely spread, the path it will take, and the location of occupants who will be in most danger if the fire does spread. Hose lines should be stretched and placed to hit the main body of fire, to cut off its spread, and to cover areas into which the fire will most likely spread.

In a large, occupied residential building, the location of the fire and direction of the most smoke indicate the greatest danger areas for occupants **Figure 4-8** . To aid in the primary search and the evacuation of victims, open stairways must be protected. The fire must be driven away from them or knocked down if a stairway is already on fire. Corridors, too, must be completely controlled, both as paths of safety and to keep the fire, smoke, and gases from penetrating into rooms or apartments.

Most victims of fire are overcome by carbon monoxide gas rather than by burns; therefore, while stairways and corridors are being controlled, every effort should be made to advance hose lines into the upper areas and to ventilate the building. A search of floors above the fire should be started as soon as possible to ensure that all occupants are located and removed from exposure to the products of combustion. Here, as in all building fires, search and rescue must be coordinated with a properly mounted attack on the fire.

Protection of occupants trapped inside the building and control of the fire depend on the number of properly sized hose lines positioned in key locations. If the fire has gained considerable headway, for example, streams from a 2½-inch

Key Points

In large, occupied residential buildings, hose lines should be placed to hit the main body of fire, to cut off its spread, and to cover areas into which the fire will most likely spread.

Key Points

To aid in the primary search and the evacuation of victims, fire must be driven away from open stairways or knocked down if a stairway is already on fire.

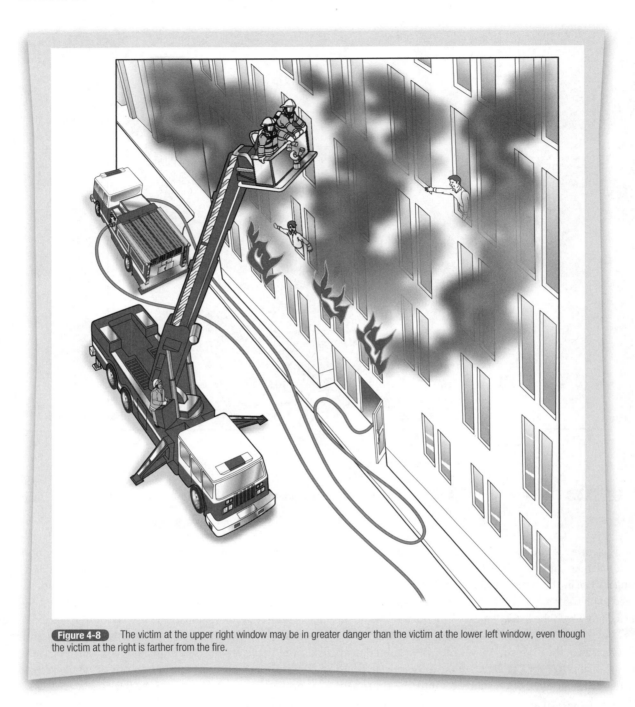

Figure 4-8   The victim at the upper right window may be in greater danger than the victim at the lower left window, even though the victim at the right is farther from the fire.

hose line should be used on the main body of fire **Figure 4-9a**, whereas 1¾-inch hose lines should be used to cut off the spread by getting around and/or above the fire **Figure 4-9b**. Small attack

lines will probably not control or extinguish a fire that has gained considerable headway. Big lines, 2½-inch attack lines must be placed into operation to control and extinguish a big fire.

Hospitals, Schools, Institutions

Fires in hospitals, schools, nursing homes, and similar institutions are attacked in essentially the same way as fires in multiple-family residences. The search and rescue problem may be compounded by the larger number of people and their age and physical condition. In particular, the first-arriving companies may be needed immediately to assist and direct people from the building.

Key Points

Protection of occupants trapped inside the building and control of the fire depend on the number of properly sized hose lines positioned in key locations.

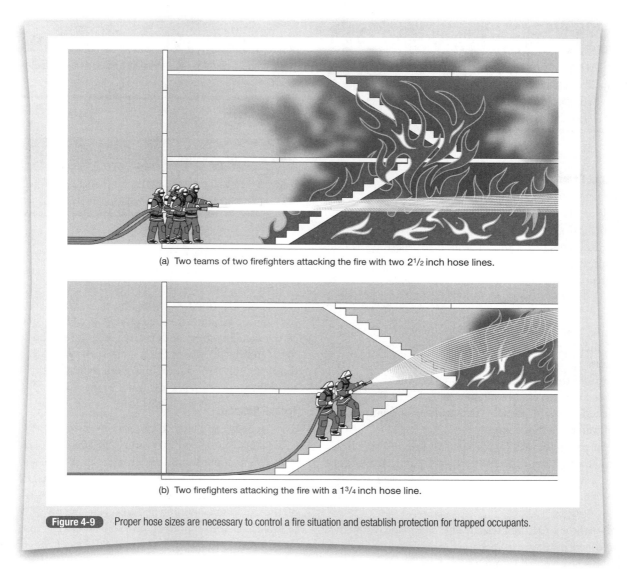

(a) Two teams of two firefighters attacking the fire with two 2½ inch hose lines.

(b) Two firefighters attacking the fire with a 1¾ inch hose line.

Figure 4-9   Proper hose sizes are necessary to control a fire situation and establish protection for trapped occupants.

Officers should not hesitate to call for help in such cases. If an engine company is performing necessary rescue duties, it must be replaced by another company, even though it may eventually be free to place its hose lines.

In all cases, it is the job of the engine company to contain the fire, open passageways, and extinguish the fire. The use of fire protection systems, if installed, such as sprinklers and standpipes, should be used to the advantage of the fire department.

Schools should be completely evacuated. Whether or not it is necessary to evacuate a hospital, nursing home or other such building depends on the type of construction, the size of

the building, and the location and severity of the fire. Patients can be moved to locations within the building that are isolated from the fire area if it is evident that the fire companies will be able to control the fire. This might be preferable to evacuation in consideration of the physical condition of patients, the continuing care that some must receive, and possibly adverse weather conditions; however, if there is any doubt regarding control of the fire, complete evacuation plans must be put into effect.

Key Points

In all cases, it is the engine company's job to contain the fire, open passageways, and extinguish the fire.

Key Points

Schools should be completely evacuated. The evacuation of hospitals, nursing homes, or other such buildings depends on the type of construction, the size of the building, and the location and severity of the fire.

Key Points

Fire companies should be prepared to use 2½-inch handlines at any fire in a fire-resistant building. The heat buildup near the fire can become so intense that smaller lines may not deliver water fast enough to absorb the heat and control the fire.

Fire-Resistant Construction

Fire-resistant structures, built to resist the spread of fire, also tend to hold in the heat rather than let it escape. Fire in such buildings forces fire fighters to combat excessive concentrations of heat as they attack the fire or search for victims. These buildings also will hold the large volumes of smoke given off by their burning contents. The smoke can quickly overcome occupants, who may collapse in just about any location inside.

In fire-resistant buildings, stairways are enclosed; smoke doors and fire doors are used to divide long corridors into smaller sections, and generally, building materials are less apt to burn. Because there is usually less spread of fire, fewer lines may be needed to contain and extinguish it; however, the intense heat associated with fires in these structures may require large volumes of water to assure control.

Fire companies should be prepared to use 2½-inch hose lines at any fire in a fire-resistant building. The heat buildup near the fire can become so intense that smaller hose lines may not deliver water fast enough to absorb the heat and control the fire.

In fires in these structures, rescues could be made within the apartment, office, or room in which the fire originated; however, the fire might have spread into the corridors, or smoke might have begun to spread through some of the floors. Frightened occupants, who would have been better off to stay in their apartments or offices, might attempt to reach an exit and collapse en route; therefore, engine company members advancing on the fire must search every area through which they pass. These areas include lobbies, elevator alcoves, corridors, and the like. Ladder and rescue companies should be assigned to thorough search operations.

Search

A thorough, planned primary search should be conducted at every fire where it is safe for fire fighters to enter the building. Moreover, all fire fighters, no matter what type of company to which they are assigned, should be able to conduct a primary search if needed. All fire fighters should realize that the safety of each member performing a primary search is their responsibility, and they must constantly be aware of the presence and position

Key Points

A thorough, planned primary search should be conducted at every fire.

Key Points

All fire fighters, no matter what type of company to which they are assigned, should be able to conduct a primary search if needed.

of each member. Always assume that there may be occupants in a building, until a search is completed, even in vacant buildings.

Because this task is so much an overall responsibility, it should be performed according to a SOG. This guideline should provide for a safe and efficient search, which is coordinated by the incident commander. The guideline should require that the search begin where there is most danger to occupants. It should be simple and straightforward so that one fire fighter can substitute for another at any point in the search.

Typical Search

Consider a fire in a large kitchen/dining area on the first floor of a two-story, single-family dwelling. Engine companies arriving at the scene would immediately size up the fire situation and obtain a continuous water supply. Normal procedures would have attack lines advanced into the house and positioned to cut off the fire from extending and to attack directly the main body of fire. Hose lines should be positioned between the occupants and the fire. Occupants closest to the fire are in the most danger whether they are on the fire floor or the floor above.

Search begins immediately. Fire fighters on the hose lines, by getting low to the floor, can probably see some clear area above the floor as they search for victims near the fire. The stairway and the upper floor will be full of smoke and hot gases. An immediate attempt should be made to advance fire fighters to the upper floor. If the area is tenable, they can begin searching for victims. If the area is untenable because of the intense heat, ventilation should be conducted from the outside **Figure 4-10** . The attack should be coordinated, with fire fighters inside the building attacking the fire and performing the primary search and fire fighters outside the building performing ventilation. Being able to perform proper ventilation procedures will depend on the number of fire fighters available for this task. As soon as the second floor is tenable, the search can begin.

Many fire departments are using positive pressure ventilation to force fresh air into a building. A powerful fan can displace a contaminated atmosphere by pushing heat and the products of combustion out of the building. This allows the search and attack

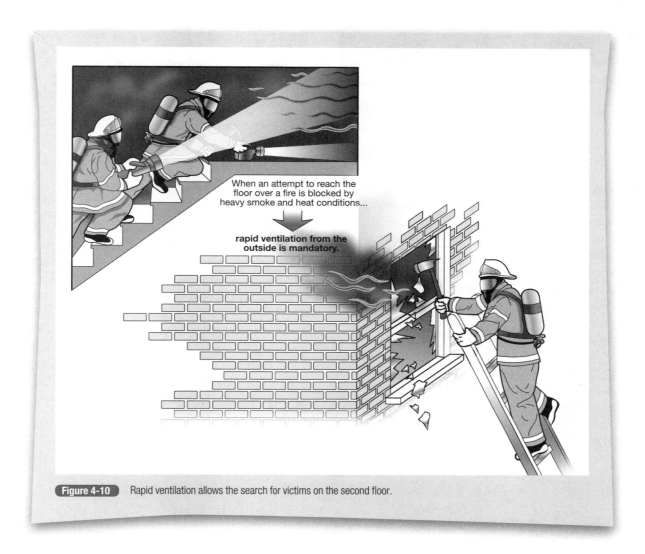

Figure 4-10   Rapid ventilation allows the search for victims on the second floor.

team entry into the structure in front of this fresh air to conduct a primary search. There are advantages as well as disadvantages in the use of positive pressure ventilation. Fire fighters must be trained to recognize when this tactic can be used safely and effectively.

Value of the Standard Search Procedure

The important point is that everyone at the fire ground knows that an immediate attempt will be made to reach the second floor. Fire fighters on the hose lines and those performing ventilation tasks must be aware that a primary search is being conducted on the upper floor.

Key Points

An immediate attempt should be made to advance fire fighters to the upper floor. If the area is tenable, they can begin searching for victims. If the area is untenable because of the intense heat, ventilation should be conducted from the outside.

Suppose now that fire fighters on the hose lines must retreat from the building because of deteriorating conditions. Following SOGs, they know they must notify fire fighters conducting the primary search on the second floor. There should be no hesitation in the notification. Those assigned to laddering and ventilation duties, knowing that a search is being conducted, should place ladders to the second floor as additional exits for search personnel and possible victims. Hose lines should be placed to assist search personnel who may be exiting the building. The rapid intervention crew should be standing by to assist in the rescue of fellow fire fighters if the need arises.

Standard Search Patterns

As soon as the second floor is tenable, the primary search will begin. Following department SOGs, the first area to be searched is directly over the fire. After reaching the top of the stairway, fire fighters will turn in one direction or the other to get to the room over the fire. They will begin the search by turning either right or left into that room. This turn sets up the basic search pattern for the entire floor. As the search proceeds, fire fighters performing the primary search will keep turning in the original direction as they go in and out of rooms.

Thus, if the first turn into the room over the fire is a right-hand turn, then all turns into subsequent rooms on the floor will be right-hand turns **Figure 4-11**. If the first turn into the room over the fire is a left-hand turn, then all turns into subsequent rooms will be left-hand turns.

For example, assume that fire fighters engaged in the search turn right at the top of the stairs and turn right again to move down the hallway to get to the room over the fire. Kicking off the search pattern for the floor, the fire fighters turn right to get into the room over the fire, as shown in Figure 4-11. When they have finished searching that room, they turn right out of the room and begin the search of the hallway until they reach the next room to be searched. Then, according to the search pattern, the fire fighters turn right again to enter a room and, after coming out of that room, again turn right and move along the hallway to the next room. The right-turning path is continued until the

fire fighters have worked all the way around the hall and are back at the stairs.

If the fire fighters need to turn back, reversing direction and following a left-left pattern will lead them back to the original starting point. Fire fighters should try to keep in mind where they are in the building and in which direction they are heading. Landmarks inside the search area should be noted, including windows or other areas that may be used for escape. These landmarks can be used to guide a fire fighter out of the building. Search rope has been used by fire fighters for many years. A search rope (or taglines) of a predetermined length will allow a fire fighter to tie off and then proceed a prescribed distance. By following the rope back, they have ensured that they will return to the original starting point.

What to Check

The corridor or hallway should be checked thoroughly, as well as the open areas of each room. In addition, closets; spaces

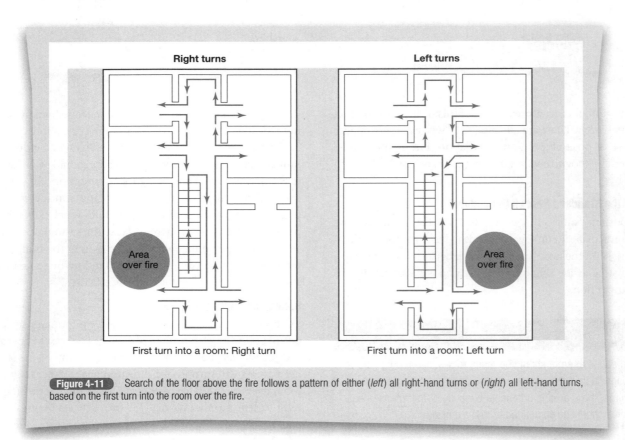

Right turns **Left turns**

Area over fire

Area over fire

First turn into a room: Right turn First turn into a room: Left turn

Figure 4-11 Search of the floor above the fire follows a pattern of either (*left*) all right-hand turns or (*right*) all left-hand turns, based on the first turn into the room over the fire.

behind large chairs, furniture, and large appliances; areas beside and under beds; bathrooms, including tubs and showers; and any other area in which a person would seek refuge should be checked. People often seek protection in such places. Everyone engaged in the search should carry a fire ax, halligan tool, or claw tool with which to probe areas and reach out into open areas of the room. These tools are also be used for forcible entry and ventilation tasks.

Rooms that the fire has entered should also be searched if possible, then the door should be shut to isolate the fire. Standard guidelines should include an engine company advancing a hose line over the fire for just this situation. The windows of rooms that are not involved with fire and hallway windows should be opened or removed in order to vent as much of the upper story as possible.

Indicating that a Room Has Been Searched

A standard way of indicating that a room has been searched should always be used so that there is no duplication of effort, at least in the initial search operation. One effective method is to place a piece of light furniture in the doorway **Figure 4-12** . A chair, footstool, end table, lamp table, or anything else that can be quickly dragged or carried into position will do. The piece should be set on its side so its legs are pointing out of the room. Other search personnel can quickly do and easily recognize this. Also, there is little chance that a piece of furniture could be knocked into such a position accidentally. Furthermore, a fire fighter does not have to enter the room to find out that it has already been searched. As an alternative, some departments have had success with placing tags or straps on the doors or doorknobs of inspected rooms, apartments, and offices. These devices can be purchased from a vendor or fabricated by department members. You can also use chalk to mark the door to indicate that the room has been searched.

Figure 4-12   Placing a chair in a doorway is one way of indicating that a room has been searched.

Other Structures

The standard search procedures and search pattern described here apply to other buildings as well as to the two-story house. The same search pattern can be used in a single-story dwelling, an apartment house, or an office building. After they turn off a corridor, into an apartment or office, for instance, fire fighters engaged in the search should follow the same pattern within the unit. When they leave the unit, they should retain the pattern in the corridor and when they enter the next unit.

More fire fighters will be needed to search larger and/or more complicated structures. Unfortunately, if there is not an adequate firefighting force on hand that is committed to the primary search, it may become a difficult task to complete in a reasonable amount of time. The need for additional fire fighters and/or adjustments in strategic goals will be the incident commander's decision.

Key Points

Any area in which a person would seek refuge should be checked thoroughly, including the corridor or hallway; the open areas of each room; closets; spaces behind large furniture and large appliances; areas beside and under beds; and bathrooms, including tubs and showers.

Key Points

A standard way of indicating that a room has been searched should always be used so there is no duplication of effort, at least in the initial search operation.

Chief Concepts

- There is much more to rescue than carrying victims from a fire building. Rescue begins well before the alarm and culminates in the standard search and rescue procedure at the fire scene.
- Many rescue connected duties must be performed by many fire fighters working together.
- The primary function of the engine company in a rescue situation is to support the primary search.
- Engine companies are responsible for the proper placement of hose lines between the fire and victims.
- Engine companies are also responsible for protecting fire fighters conducting the search. They must also obtain water and place streams to control the fire.

Key Term

Rapid intervention team (RIT): A minimum of two fully equipped personnel on site, in a ready state, for immediate rescue of injured or trapped fire fighters.

Fire Fighter in Action

1. The greatest number of fire deaths occur in _____ occupancies.
   - **a.** Educational
   - **b.** Health care
   - **c.** Industrial
   - **d.** Residential

2. Preparation for rescue begins
   - **a.** After arrival, but after a preliminary sizeup
   - **b.** After arrival after an in-depth sizeup
   - **c.** When the alarm is received, by considering time and weather factors
   - **d.** Well before the incident by preincident planning

3. Fires in residential properties
   - **a.** Always include the possibility of a rescue situation
   - **b.** Present a rescue problem at night and on weekends
   - **c.** Present a rescue problem during the afternoon, evening hours, and on weekends
   - **d.** Seldom require actual rescues. With the widespread use of smoke detectors in these properties, occupants evacuate before arrival of the fire department.

4. Victims at windows indicate
   - **a.** That all occupants are probably at positions visible from the exterior; thus, rescue efforts should concentrate on quickly removing all visible occupants
   - **b.** That all occupants are aware of the fire
   - **c.** That other occupants may still be trapped inside
   - **d.** Victims at windows is not a good indicator of anything but the obvious need to rescue visible occupants.

6. The first-arriving engine company confronted with a working structure fire with possible victims should usually
   - **a.** Conduct an immediate primary search
   - **b.** Conduct a vent, entry, search procedure
   - **c.** Attack the fire
   - **d.** Any of the above would be considered a usual first action.

7. A rapid intervention team (RIT) must be available
   - **a.** From the initial stages of an incident to its conclusion
   - **b.** After a second crew enters the hazard area
   - **c.** As soon as on-scene staffing permits
   - **d.** Whenever the incident commander determines fire fighters are at risk

8. A rapid intervention team (RIT) consists of at least _____ fire fighters in full PPE.
   - **a.** 1
   - **b.** 2
   - **c.** 3
   - **d.** 4

9. The search for victims is normally the function of a(n)
   - **a.** Engine company
   - **b.** Rescue or ladder company
   - **c.** Special rescue team of fire fighters formed into a task force at the incident scene
   - **d.** All of the above

10. Most fire victims are overcome by _____.
    - **a.** Burns
    - **b.** Carbon monoxide
    - **c.** Carbon dioxide
    - **d.** Carbon particulates entrained in the smoke

11. When confronted with a school fire, the decision to evacuate is based on
    - **a.** The type of construction
    - **b.** Size of the building
    - **c.** Location and severity of the fire
    - **d.** Schools should be completely evacuated, regardless of these factors.

12. You are the officer of the second-arriving company for a residential fire on the first floor that is being attacked by the first-arriving engine company. The incident commander assigns your company to search the second floor. You should
    - **a.** Ascend the main stairway to the second floor. Turn in a direction that places you directly over the fire. Begin searching there, and continue to search in the direction you turned at the top of the stairway.
    - **b.** Ascend the main stairway to the second floor. Turn in either direction and begin searching. Continue to search in the direction you turned at the top of the stairway.
    - **c.** Ascend the main stairway to the second floor. Turn in a direction that places you directly over the fire. Begin searching there, and continue to search using either a right- or left-hand search pattern.
    - **d.** Ascend the main stairway to the second floor. Turn in either direction and begin searching. Continue to search using either a left- or right-hand search pattern.

13. Members of the search team should carry a fire ax, halligan tool, or claw tool. These tools are used by the search team to
    - **a.** Force entry into rooms
    - **b.** Ventilate
    - **c.** Probe areas when conducting a search
    - **d.** All of the above

14. You are assigned to search the second floor for a residential fire that originated on the first floor. As you search, you open a bedroom door and encounter heavy smoke and fire in that room. Fire conditions make the room untenable and the search of that room is thus delayed. You should
    - **a.** Delay the primary search of the remainder of the second floor until a hose line is in position to attack the fire.
    - **b.** Move to the exterior to vent the involved room.
    - **c.** Close the door to isolate the fire, and continue searching other rooms on the second floor if possible.
    - **d.** Immediately notify command, gather your crew, and conduct an emergency retreat.

Water Supply

Learning Objectives

- Recognize the four possible sources of water at the fire ground.

- Understand the components and function of a pumper, whose primary purpose is to combat structural and associated fires.

- Understand the various factors that influence the flow of water through the fire hose.

- Consider the type of hose, diameter, length, and carrying capacity used on the fire ground.

- Examine the reasons for a pumper relay and the procedures needed to be taken for proper operations.

Obtaining an adequate and continuous water supply on the fire ground is a basic firefighting task. Without water, fire fighters would be unable to attack, control, and extinguish a fire. Without a continuous supply of water, members engaged in firefighting operations could be seriously injured or killed. Fire fighters depend on this resource to be available when needed and in sufficient quantities to accomplish their objective.

To achieve this goal, water supply operations provided by engine companies must be carried out safely. Standard operating guidelines as well as approved safe work practices should be followed. Water sources, such as hydrants systems or static water sources, pumpers, and mobile water supply apparatus, as well as the supply hose, should all be addressed when considering safety issues. The following tasks have inherent risks regarding safety:

- Laying of supply lines
- Connecting a pumper to a water supply
- Supplying water to the fire ground
- Obtaining water from a static water source
- Tanker shuttle operations
- Relay pumping operations

Fire fighters must pay attention when performing these tasks and recognize the fundamental danger of an activity that may appear to be routine but could cause serious injury or death. An example would be driving a pumper or mobile water supply apparatus during a tanker shuttle operation. The driver/operator is responsible for getting the apparatus to the scene safely following state law and department standard operating guidelines. Unfortunately, there are far too many accidents each year in which the apparatus is involved in an accident because of inattention or poor judgment by fire department personnel. Fire fighters must consider safety as a top priority in all aspects of firefighting including those involving water supply operations.

Engine companies exist because of their ability to supply water at the fire ground. Their apparatus is designed and equipped for that job. Personnel are trained to provide a continuous, uninterrupted supply of water quickly and efficiently.

Chapter 2 explained how the first action of an engine company arriving at the fire ground is to obtain water for firefighting and rescue. Where the initial action is laying supply lines, this operation should involve as few fire fighters as possible. The reason for this is obvious: As many fire fighters as possible should be free for initial fire attack and rescue.

Three factors influence the movement of water at fires:

- Water source
- Pumper
- Hose

This chapter discusses each of these in turn, always with the objective of obtaining maximum utilization with minimum personnel. Water delivery limitations are noted, as well as some operations that can be effective in spite of these limitations. Also included is a discussion of the relay operations necessitated by long supply line lays and a section on pump performance.

Water Source

There are four possible sources of water at the fire ground:

- Water main systems
- Static water sources
- Apparatus water tanks
- Mobile water supply apparatus

Water Main Systems

Water main systems are the major source of water for many fire departments; however, many fire departments may have little control over them. An engine company must accept what is available from a water main system and must make the best use of it. It is important, therefore, to know the flow rates of the system, of the parts of the system servicing particular areas, and of individual hydrants. Flow-test data can be used to determine these rates. The fire department should either run its own tests or be a part of the process if testing is done by another agency.

If a hydrant or group of hydrants provides a less-than-adequate supply of water, additional sources must be used. Sources used to augment the basic water supply should be identified by preincident planning before a major incident occurs. These might simply consist of additional hose lays to more distant hydrants known to have ample water, or they can include the use of static water sources and/or pumper relays. Areas with dead-end mains and small mains are especially vulnerable to water supply problems. The use of a large-diameter hose (LDH), if not already routinely used, should be considered in these areas.

Among the factors that affect the flow rate of a water main system are the sizes of mains, the capacities and locations of elevated reservoirs, and the capacities of pumps if used in the system. These are all design features, difficult and costly to change once a system has been established.

Reading Pressure Levels

Static pressure is the pressure of the hydrant water at rest—that is, with the hydrant open to the pump and no water flowing through the pump. **Residual pressure** is the pressure in the hydrant with water flowing from the hydrant through the pump. Both are

Key Points

Water main systems are the major source of water for many fire departments; however, many fire departments may have little control over them. An engine company must accept what is available from a water main system and must make the best use of it.

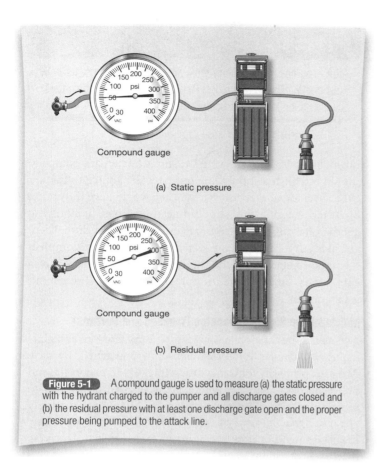

Figure 5-1 A compound gauge is used to measure (a) the static pressure with the hydrant charged to the pumper and all discharge gates closed and (b) the residual pressure with at least one discharge gate open and the proper pressure being pumped to the attack line.

read on the compound (intake) gauge located on the pump panel **Figure 5-1** . Together, they can be used by the pump operator to obtain an accurate indication of the available water supply from the hydrant. At the beginning of the operation, the pump operator reads the compound gauge with the hydrant open and all discharge gates closed. This is the static pressure. With one attack line charged to the proper discharge pressure, the compound gauge is read again. This is the residual pressure.

Static pressure by itself provides no indication of available water. It is the difference between static and residual pressure that gives you the true measurement.

Estimating Attack Lines

The difference in the readings allows the pump operator to estimate how many more attack lines can be supplied by the water main system. As shown in **Figure 5-2a** , a drop of about 5% from the static pressure to the residual pressure indicates that three equal parts equivalent to the amount being delivered can be supplied by the water main system. **Figure 5-2b** shows a drop of about 10%, which indicates that two more equal parts can be supplied. **Figure 5-2c** shows a drop of about 20%, indicating that only one more equal part can be delivered from the water main system.

Even after all attack lines are charged, the operator must continually watch the compound gauge closely. This is to assure that immediate action can be taken if other pumpers operating nearby cause the residual pressure to decrease below the minimum operating level.

Static Water Sources

When drafting is the only way to assure an adequate water supply, preincident planning is a must. Locations of static water sources, the volume of water available (especially from ponds and pools), and their distances from the structures they could serve should be known in advance of a fire. Alternative and supplementary static water sources should be determined as part of preincident planning.

Key Points

The difference in the static pressure and the residual pressure readings allows the pump operator to estimate how many more attack lines can be supplied by the water main system.

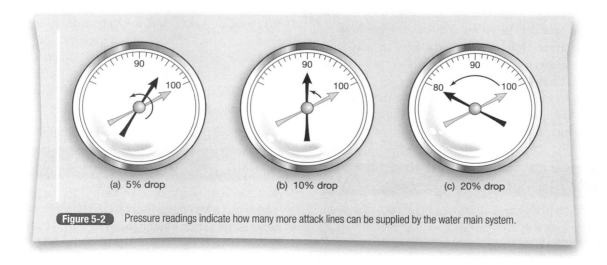

Figure 5-2 Pressure readings indicate how many more attack lines can be supplied by the water main system.

Often a static water source or hydrant is located some distance from the fire structure. A pumper relay may be necessary to move water efficiently from such a source to the fire ground **Figure 5-3**. Water for the initial attack will most likely come from a pumper's water tank and/or mobile water supply apparatus.

Apparatus Water Tanks

If possible, given the fire situation and the initial hose lay, the first-arriving pumper may begin fire attack with water from its water tank. This procedure eliminates the need to wait until supply lines are charged or a relay is set up. Water is delivered to the fire immediately.

After a pumper is supplying tank water to attack lines, its tank must not be allowed to run dry. The consequences to fire fighters, occupants, and the fire structure could be disastrous.

Key Points

When drafting is the only way to assure an adequate water supply, preincident planning is a must.

Key Points

After a pumper is supplying tank water to attack lines, its tank must not be allowed to run dry. The consequences to fire fighters, occupants, and the fire structure could be disastrous.

In smoke and/or fire showing situations, the engine company should establish a supply line from a water source to the fire either by a forward or reverse lay. They would have the option of either charging this supply line themselves or notifying another arriving engine company to perform that duty. There are many options that fire departments use to accomplish this task dependent on their individual circumstances, but a continuous, uninterrupted supply of water to the fire ground is paramount.

Mobile Water Supply Apparatus (Tankers and Tenders)

Two main types of mobile water supply apparatus are available: tankers and tenders. A mobile water supply apparatus is a vehicle designed primarily for transporting (pickup, transporting, and delivering) water to fire emergency scenes to be applied to other vehicles or pumping equipment. Mobile water supply apparatus generally carry 1,000 to 3,500 gallons of water. Many fire departments operate tractor drawn tanks capable of carrying thousands of gallons of water.

Mobile water supply apparatus may or may not be equipped with a fire pump. Some vehicles are equipped with a permanently mounted fire pump of 750 gpm or more, whereas others may have a smaller capacity pump or a power take off (PTO) drive system. Pump capacities are predicated on a fire department's needs.

Vehicles carrying 1,000 gallons of water or more with a fire pump are generally referred to as tankers. Pumper-tankers have a standard fire pump and a large water tank. This type of pumper should be used for fire attack or pumping water from a water source to pumpers on the fire ground. It should not be used to shuttle water from a water source to the fire ground. Vehicles carrying 1,000 gallons of water or more without a fire pump or using a small PTO-driven pump are called tenders. These vehicles are specifically used to carry water to a fire.

A mobile water supply apparatus can be used for the initial water supply until a continuous water source is obtained from

Key Points

A mobile water supply apparatus can be used for the initial water supply until a continuous water source is obtained from a static water supply or from a hydrant system.

a static water supply or from a hydrant system. In this case, the mobile water supply apparatus would provide a pumper directly by supplementing tank water until the continuous water source is obtained through the use of supply lines.

In rural settings, a tanker shuttle operation may be required to accomplish this task. When mobile water supply apparatus are used for developing a continuous water supply, they can either supply the pumpers directly or dump their water into a portable water tank from which the pumpers draft. The number of mobile water supply apparatus needed for a tanker shuttle operation will depend on the distance from the fire to the water source, the size of the fire, how long it takes to fill the water tank at the source, and how long it takes to dump the water at the fire ground, as well as traffic and road conditions.

Large modern dump valves allow water to be offloaded with minimal time loss enabling the truck to return quickly to reload **Figure 5-4** . Additional methods might be desired to improve the offloading rate of gravity dumps. These methods include a jet assist or a pneumatic pump. Basically, a jet is a pressurized water stream used to increase the velocity of a larger volume of water that is flowing by gravity through a given size dump valve. A pneumatic system can be used to pressurize a tank and assist in expelling the water. To make the overall tanker shuttle operation more effective, a pumper or pumpers should be located at the water source to fill the mobile water supply apparatus **Figure 5-5** . This method is much faster than having these vehicles fill from small pumps or portable pumps.

In the following scenario, two pumpers and mobile water supply apparatus are used. The first pumper arrives at the fire scene and begins the initial attack with water from its water tank. The second pumper arrives and supplies the first pumper with water from its water tank. The second pumper also sets up to receive water from a mobile water supply apparatus. This can be accomplished by receiving water directly into the pump or setting up a portable water tank for drafting operations. The mobile water supply apparatus arrives and provides water to the second pumper, which in turn pumps water to the first pumper. This assures that the initial attack will be maintained, that there is minimal movement of apparatus on the fire ground, and that the water supply is located at a relatively safe and accessible distance from the fire.

After the second pumper is set up, fire fighters can assist members of the first pumper in operating attack lines or performing other firefighting tasks. Furthermore, the second

Figure 5-4   A dump valve discharges water into a portable water tank.

Figure 5-5   A pumper drafts water from a portable water tank.

Key Points

The number of attack lines to be placed in service will be determined by the supply of water that can be maintained by the tankers.

pumper can then be set up to supply another pumper located on the fire ground. The number of attack lines to be placed in service will be determined by the supply of water that can be maintained by the mobile water supply apparatus **Figure 5-6**.

Preincident planning is the key to establishing standard operating guidelines affecting water supply requirements. A fire department must recognize its capability, as well as those of other agencies, and be able to work within its parameters to provide a continuous water supply of adequate volume for a particular incident.

Pumper

A **pumper** is a piece of fire apparatus with a permanently mounted fire pump of at least 750 gpm (3,000 L/min capacity), a water tank, and a hose body whose primary purpose is to combat structural and associated fires.

The water delivery capacity of a pumper is limited by the capacity of the pump and by the number of suction intakes and discharges with which it is equipped. In addition to these mechanical factors, operation of the pumper is limited by engine speed and residual pressure.

Rated Capacity

The flow rate of which the pump manufacturer certifies compliance of the pump when it is new is known as rated capacity. The pumping system provided must be capable of delivering the following:

- One hundred percent of rated capacity at 150 psi (1000 kPa) net pump pressure.
- Seventy percent of rated capacity at 200 psi (1400 kPa) net pump pressure.

Key Points

The water delivery capacity of a pumper is limited by the capacity of the pump and by the number of suction intakes and discharges with which it is equipped. Operation of the pumper is limited by engine speed and residual pressure.

Key Points

The pump must have at least the number of intakes required to match the appropriate arrangement for the rated capacity of the pump, and the required intakes must be at least equal in size to the size of the suction lines for that arrangement.

- Fifty percent of rated capacity at 250 psi (1700 kPa) net pump pressure.

Pumpers are rated from draft. Thus, a 1,250 gpm pumper can draft and discharge 1,250 gallons of water per minute from a static water source. Pumpers will exceed their rated capacity if they receive water under positive pressure; thus, if the pumper is at the fire ground and is being supplied by another pumper at a hydrant or directly by the hydrant and if the hydrant delivers sufficient water, it will be able to pump more than 1,250 gpm. The limiting factors would then include the number and sizes of intakes and discharges.

Intakes

The pump must have at least the number of intakes required to match the appropriate arrangement for the rated capacity of the pump, and the required intakes must be at least equal in size to the size of the suction lines for that arrangement. Intakes must have male NH threads if the apparatus is to be used in the United States.

Suction intakes are usually located on the sides, front, or rear of the pumper. When a pumper is working directly from a hydrant, the common procedure is to hook up with standard soft suction hose, generally a short length (10 to 20 feet) of an LDH. This supply line is connected to one of the suction intakes.

When drafting, it should be noted that the side intakes on the pumper are closest to the pump itself. Intakes at the front or rear of the apparatus, or otherwise specially situated, might not allow drafting rated capacities at rated pressures due to the distance away from the pump.

When a pumper stretches supply hose to the fire using a forward lay, it is usually done with an LDH (4 to 5 inches). Many fire departments now find it useful to use large-diameter supply hose to move water effectively from the source to the fire scene **Figure 5-7**. The department should evaluate its needs

Key Points

When a pumper stretches supply hose to the fire using a forward lay, it is usually done with an LDH (4 to 5 inches).

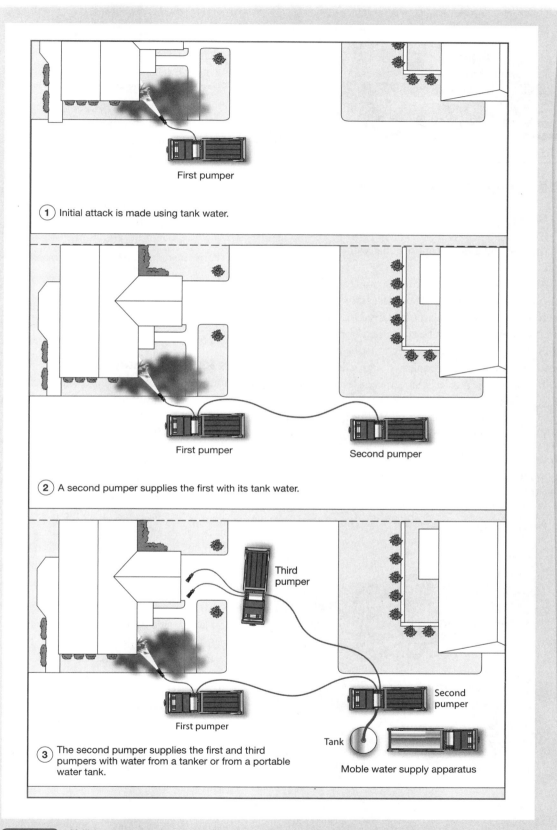

1. Initial attack is made using tank water.

2. A second pumper supplies the first with its tank water.

3. The second pumper supplies the first and third pumpers with water from a tanker or from a portable water tank.

Figure 5-6   Multiple pumpers supply water at a fire.

Figure 5-7   A large-diameter supply hose is often used to move water to the fire scene.

and choose the size and amount of hose that will best support its operation.

If the hydrant does not have a steamer connection but only 2½-inch discharge openings, it will be necessary to hook up with 2½-inch or larger hose. At least one valved intake must be provided on the pumper that can be controlled from the pump operator's position. The valve and piping must be a minimum of 2½ inches. The intake must be equipped with a female swivel coupling with NH threads. A gated suction siamese could also be used **Figure 5-8** . This adapter is placed on a suction intake of the pumper and allows for two additional supply lines, a minimum of a 2½-inch hose, to feed the pumper.

Discharges

Discharge outlets of 2½ inches or larger must be provided to discharge the rated capacity of the pump. All 2½-inch or

Figure 5-8   A gated suction siamese is used to allow water to be received from two different supply lines.

larger diameter outlets must be equipped with male NH treads. The usual practice is to install one 2½-inch discharge for each 250 gpm of rated capacity.

Discharges may be located on either side of the pumper or in the front or rear. They may be capped, as at the pump panel, or they may be preconnected to 1¾- or 2½-inch attack lines. A discharge may feed a master stream appliance located on top of the pumper.

Many fire departments miss the opportunity to get the most from a new pumper. They order a large, powerful diesel engine for road performance but do not make full use of the engine in driving the pump. Pumps should take advantage of their full pumping capacity. Adequate discharge outlets should be provided on pumpers to complement their rated capacity.

Pump Speed and Capacity

Although a pumper can deliver more than its rated capacity under certain conditions, there is a limit to how far it can be pushed. This limit may be imposed either by pump speed or true pump capacity, depending on the situation.

The pump must not be operated above 80% of its rated peak speed for any length of time. If it is, the engine, drive train, or pump may be damaged. The pump operator can tell when the pump is operating at its peak capacity by watching the discharge pressure gauge. If an increase in engine pump speed (rpm) is not accompanied by an increase in discharge pressure, the pump is moving as much water as it can. It has reached its true capacity or the limit of the water supply **Figure 5-9** .

Residual Pressure

Residual pressure, as noted earlier, is the pressure remaining on the intake side of the pump while water is flowing through the pump. It is, in a sense, a measure of the reserve capacity of the hydrant. When it reaches zero, there is no flow because there is no water left.

For this reason, the pump operator must continually monitor the pumper's residual pressure. This is especially im-

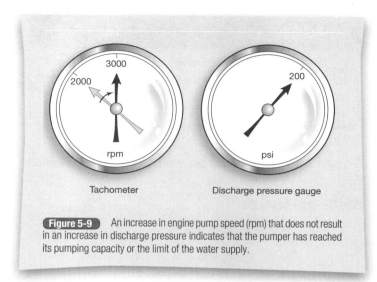

Figure 5-9 An increase in engine pump speed (rpm) that does not result in an increase in discharge pressure indicates that the pumper has reached its pumping capacity or the limit of the water supply.

portant when several pumpers are drawing water from the same hydrant system. The more pumpers drawing water, the more acute a water supply problem can be and the more quickly it can develop.

Pump operators should maintain a residual pressure of 20 to 25 pounds per square inch (psi) or more if at all possible. The residual pressure should not be allowed to drop below 10 psi, except under extreme conditions. When it looks as if this might happen, the pump operator must notify the company officer immediately, while attempting to maintain at least enough residual pressure to avoid losing water completely. The proper procedure is to lower the discharge pressure of the pump. This will reduce the pressure on all the hose lines operating off the pump. For safety reasons, the pump operator must never shut down a hose line to reduce the flow without the knowledge and consent by personnel operating that hose line. If action is required to limit the flow of water to increase residual pressure, the IC should determine what actions should be taken to correct the problem. Increasing residual pressure could be accomplished by shutting down an unnecessary hose line(s), half gating hose lines, or reducing tip sizes to decrease flows. A supplemental supply line should be placed in service as quickly as possible to reestablish an adequate water supply to the pumper, which will increase the residual pressure.

The residual pressure at a hydrant is decreased by increases in the pumper discharge pressure, discharge volume (gpm), and

speed (rpm). Each of these has the effect of increasing the rate at which water is used from the hydrant. The more water taken from the hydrant, the less water remains as potential capacity. One way in which the engine company can increase the efficiency of its operations is to minimize friction losses in the supply hose and attack lines.

The friction loss characteristics of fire hose are an important consideration in the selection of hose. Friction loss varies considerably depending on the construction and design of the hose, the roughness of the lining, and its internal diameter. The type of couplings can also affect the friction loss.

Hose

The third factor that influences the movement of water at fires is the hose. Fire hose is the flexible conduit used to convey water. There must be enough hose lines, with sufficiently large diameters, to carry the water. Because of friction loss in the hose, the diameter of the hose and the length of the hose line from the pumper to the fire stream directly affect the water-moving capability of the system. The larger the diameter of the hose, the less friction loss there is for a given flow rate. Conversely greater the flow rate in a given size hose, the greater the friction loss is. **Figure 5-10** illustrates some various fire hose sizes.

To provide proper nozzle pressures, pumpers must overcome the friction losses. Effective use of hose lines can reduce required

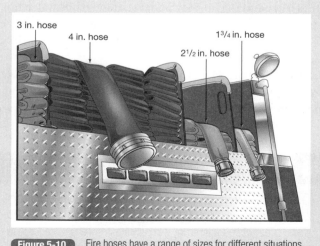

Figure 5-10 Fire hoses have a range of sizes for different situations.

Key Points

Because of friction loss in the hose, the diameter of the hose and the length of the hose line from the pumper to the fire stream directly affect the water-moving capability of the system.

Key Points

An attack hose is designed to convey water to hose line nozzles, distributor nozzles, master stream appliances, portable hydrants, manifolds, standpipe and sprinkler systems, and pumps used by the fire department.

Key Points

Supply hose is designed for moving water between a pressurized water source and a pump that is supplying attack lines.

pump pressures and permit more efficient movement of water. An attack hose is designed to be used by trained fire fighters to combat fires beyond the incipient stage. The attack hose is designed to convey water to hose line nozzles, distributor nozzles, master stream appliances, portable hydrants, manifolds, standpipe and sprinkler systems, and pumps used by the fire department. The attack hose is designed for use at operating pressures up to at least 275 psi.

The supply hose is designed for moving water between a pressurized water source and a pump that is supplying attack lines. An LDH has a diameter of 3½ inch or larger. The 4- and 5-inch LDHs are the most common sizes used in the fire service. An LDH is effective in moving large amounts of water from pumpers at a water source to other pumpers at the fire scene. An LDH also can be used by a pumper to lay a supply line from a water source to the fire scene for a continuous, uninterrupted water supply. An LDH also is used to supply attack lines, master stream appliances, portable hydrants, manifolds, and standpipe and sprinkler systems. Whenever used for these applications, a

Key Points

An LDH is effective in moving large amounts of water from pumpers at a water source to other pumpers at the fire scene. An LDH also can be used by a pumper to lay a supply line from a water source to the fire scene for a continuous, uninterrupted water supply.

Key Points

The standard hose of the fire service nationally is the 2½-inch diameter line. It can be used for attack lines; for supplying master stream appliances, standpipes, and sprinkler systems; and for water supply.

pressure relief device with a maximum setting of 200 psi should be used.

One 2½-Inch Hose Line

The standard hose of the fire service nationally is the 2½-inch diameter line. It can be used for attack lines; for supplying master stream appliances, standpipes and sprinkler systems; and for water supply. Other sizes may fill some of these needs, but not all of them; however, the 2½-inch hose is not perfect: It does have limitations, especially when used to move large quantities of water.

For example, a pumper at the fire receiving water from a hydrant or from a pumper at a hydrant might be attempting to supply two 1¾-inch attack lines along with a 2½-inch attack line equipped with spray nozzles. A flow rate of about 550 gpm would be required. In working directly from the hydrant, it must be known ahead of time that the hydrant is capable of maintaining such a flow and of providing an acceptable residual pressure at the same time. There would be a limit of an operation with a single 2½-inch hose line between two pumpers; the flow could be maintained only if the pumpers were comparatively close to each other. Thus, a 2½-inch supply line should not be used to deliver much more than 350 gpm. The friction loss in 2½-inch hose at these flows is 25 psi per 100 ft of line.

It would be far better to use two 2½-inch hose lines for supply or a single LDH. **Figure 5-11** shows the maximum available flow using 1,000 ft as a standard supply line at a pump discharge of 175 psi. In this illustration, a single 2½-inch supply line is capable of flowing only 275 gpm.

Two 2½-Inch Lines

Using two 2½-inch supply lines instead of one to move a given amount of water can reduce friction loss substantially. Suppose two 2½-inch supply lines are used to move the 500 gpm. Now, because only 250 gpm are passing through each hose line, the friction loss per 100 ft is reduced to about 15 psi.

Key Points

Using two 2½-inch supply lines instead of one to move a given amount of water can reduce friction loss substantially.

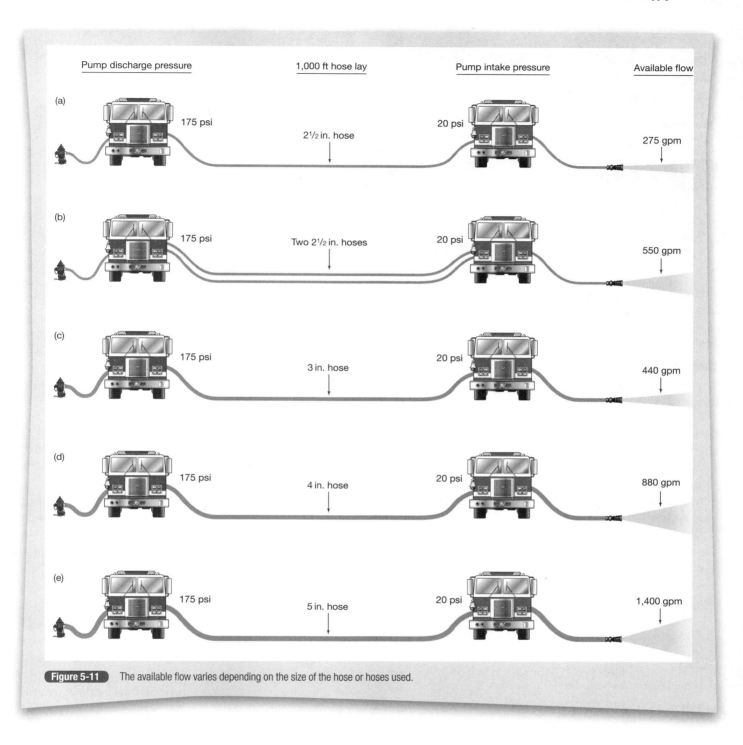

Pump discharge pressure	1,000 ft hose lay	Pump intake pressure	Available flow
(a) 175 psi	2½ in. hose	20 psi	275 gpm
(b) 175 psi	Two 2½ in. hoses	20 psi	550 gpm
(c) 175 psi	3 in. hose	20 psi	440 gpm
(d) 175 psi	4 in. hose	20 psi	880 gpm
(e) 175 psi	5 in. hose	20 psi	1,400 gpm

Figure 5-11   The available flow varies depending on the size of the hose or hoses used.

This means there is a greater likelihood of obtaining an adequate flow directly from the hydrant. In working directly from the hydrant, the use of two 2½-inch supply lines permits a more efficient operation that allows movement of more water to the pumper than the use of a single 2½-inch line. As shown in Figure 5-11, two 2½-inch supply lines are capable of flowing 550 gpm, whereas a single 2½-inch supply line is capable of flowing 275 gpm.

As long as the total flow of water and the hose line length remain the same, an increase in the number of supply lines of a given diameter will decrease the required pump pressure. Whether hose lines are carrying water to a single point or to separate positions, the friction loss is the same in each line if the flows are the same and the hose lines are the same diameter.

There is always a practical limit to the number of 2½-inch hose lines that can be connected between pumpers. The

Key Points

As long as the total flow of water and the hose line length remain the same, an increase in the number of supply lines of a given diameter will decrease the required pump pressure.

Key Points

Increasing the diameter of the hose decreases the friction loss as well as the number of hose lines needed to move a given amount of water.

alternative is to use larger diameter hose. As noted previously, increasing the diameter of the hose decreases the friction loss as well as the number of hose lines needed to move a given amount of water.

Three-Inch Hose

The next standard size of hose after 2½ inches is 3 inches in diameter. The difference between the water-carrying capabilities of the two sizes of hose may surprise anyone not familiar with the larger hose. For all practical purposes, a 3-inch line is almost as effective as two 2½ inch lines in carrying water to a single point from a hydrant or pumper, as shown in Figure 5-11. Hydraulics manuals confirm this statement. Also, for a given flow, the difference between the friction loss in a single 2½-inch hose line and that in a single 3-inch hose line is great, but the difference between the loss in two 2½-inch hose lines and that in a single 3-inch hose line is slight. The capability of moving as much water with one hose line as is usually moved with two makes the use of the 3-inch hose more practical if an LDH is available.

Three-inch hose may be ordered with 2½-inch couplings; then there is no need for adapters to hook the two sizes of hose together. Little, if any, decrease in water pressure will be noticed when these couplings are used because the increased velocity at the coupling makes up for most of the loss.

An LDH

For practical purposes, NFPA 1961, *Standard on Fire Hose*, defines an LDH as a hose with an inside diameter of 3½ inches or larger.

Key Points

The capability of moving as much water with one hose line as is usually moved with two makes the use of the 3-inch hose line more practical if an LDH is available.

Key Points

The carrying capacity of an LDH versus a standard fire hose is striking. A 4-inch hose line delivers a volume of water approximately equivalent to three and a half 2½-inch hose lines at any given pressure or distance. A 5-inch hose line is approximately equivalent to six 2½-inch hose lines.

Generally, an LDH measures 4 inches or greater. An LDH has made an impact on the fire service as a tool for moving large amounts of water over long distances. The hose is basically an aboveground water main for transporting water from a source to the fire ground. It is also used to supply attack lines, distributors, master stream appliances, portable hydrants, standpipe, and sprinkler systems Figure 5-12 . An LDH is being used by fire departments in the rural, suburban, and urban localities with great success.

An LDH provides the most efficient means of minimizing friction loss and developing the full potential of both water supplies and pumping capacities. An LDH increases the distance between the water system and the fire ground because of its lower friction loss characteristics.

The carrying capacity of an LDH versus standard fire hose is striking. A 4-inch hose line delivers a volume of water approximately equivalent to three and a half 2½-inch hose lines at any given pressure or distance. A 5-inch hose line is approximately equivalent to six 2½-inch hose lines.

As shown back in Figure 5-11, a 4-inch supply line is capable of flowing 880 gpm, whereas a 5-inch supply line is capable of flowing 1,400 gpm. For proper operation, an LDH requires fittings, valves, and adapters.

Supply Line Procedures

Training, experience, and sizeup on arrival at the fire scene will indicate what type of supply line should be used if an option is provided. If an LDH is available, it should always be used for water supply. The following little verse is easy to remember and makes a lot of sense: "When in doubt, lay it out."

There is absolutely no sense in being caught without water at a fire that "looked insignificant on arrival." A supply line should be laid from the initial pumper if there is any indication of a fire on arrival.

Key Points

Training, experience, and sizeup on arrival at the fire scene will indicate what type of supply line should be used.

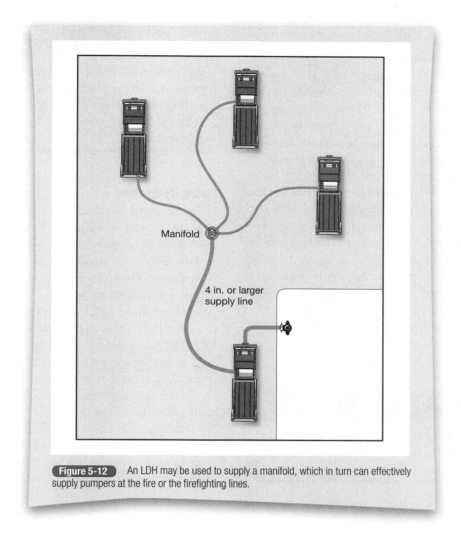

Figure 5-12 An LDH may be used to supply a manifold, which in turn can effectively supply pumpers at the fire or the firefighting lines.

A pumper approaching the fire from a water source should perform a forward lay. If the pumper arrives at the fire first, then a reverse lay, from the fire to a water source, should be initiated. The supply line should be charged, providing the pumper with a continuous water supply. If no supply line is laid and the pumper begins operation with tank water, there is a possibility that this water supply could be depleted before the fire is controlled.

The situation could become disastrous unless another pumper arrives to supplement the initial pumper's water supply. Departmental procedures may allow for an initial pumper to lay in a supply line but leave it uncharged. Another arriving pumper would be assigned to this task. If there was a delay and the supply line needed to be charged, a member of the initial pumper could be assigned this task. When water supply systems are unavailable, the task of supplying water to initial pumpers should be assigned to other pumpers or mobile water supply apparatus.

As a rule, a pumper at a hydrant should initially begin pumping to a unit at the fire at a pump pressure of 100 psi.

Under average conditions, this will provide the water and residual pressure required by the pumper at the fire. As the situation changes, or if the 100 psi is found to be insufficient, the pressure can be increased. For long lays, the initial pressure may be sufficiently higher to compensate for the increased friction loss depending on length of the lay, the carrying capacity of the supply hose, and the amount of water being used, and as always, the residual pressure must be carefully monitored and kept above the minimum.

Key Points

As a rule, a pumper at a hydrant should initially begin pumping to a unit at the fire at a pump pressure of 100 psi. Under average conditions, this will provide the water and residual pressure required by the pumper at the fire.

Pumper Relays

Excessively long lays, those exceeding about 1,000 ft (up to 2,000 ft maximum), may require relaying operations to counteract the effect of friction loss and elevation pressure. This will be true especially if an LDH is not used. These relaying operations are simple if handled properly, involving merely pumping water through two to several pumpers stationed at intervals between an adequate water source and the fire ground. The amount of pumpers and supply hose available are the only limitation to the distance water can be relayed. It is important to know before the alarm where such operations might be required, so they can be put into effect as pumpers arrive at the fire scene. Fire department personnel should be trained in the proper procedures to be applied in a relay pumping operation. This is not the time for on-the-job training. Personnel must know how to set up this system so that it can be placed in service as quickly as possible. The maximum capacity of a relay operation is determined by the smallest pumper and/or the smallest supply hose used in the relay.

Key Points

Excessively long lays, those exceeding about 1,000 ft may require relaying operations to counteract the effect of friction loss and elevation pressures.

Key Points

The actual positions taken by pumpers in the relay line and the movements involved in getting to these positions depend on the relative locations of the fire and the water source, as well as the availability of roadways and room in which to maneuver pumpers at the fire.

Setting Up the Relay

The actual positions taken by pumpers in the relay line and the movements involved in getting to these positions depend on the relative locations of the fire and the water source, as well as the availability of roadways and room in which to maneuver pumpers at the fire. There are two ways in which most relay pumping operations are set up.

In the first method, the first-arriving pumper is positioned at the fire and its crew begins attacking the fire with tank water **Figure 5-13**. Additional responding pumpers lay supply lines from the first pumper toward the water supply. Each pumper takes up its position in the relay line when it has laid its supply line.

In the second method, the first-arriving pumper lays its supply line from the property entrance to the fire building and then begins initial attack **Figure 5-14**. Additional responding pumpers lay supply lines from the first pumper's supply line toward the water supply. Each pumper takes up its position in the relay line when it has laid its supply line.

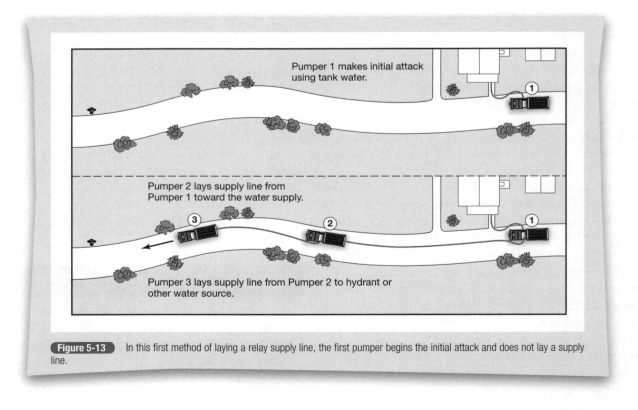

Pumper 1 makes initial attack using tank water.

Pumper 2 lays supply line from Pumper 1 toward the water supply.

Pumper 3 lays supply line from Pumper 2 to hydrant or other water source.

Figure 5-13 In this first method of laying a relay supply line, the first pumper begins the initial attack and does not lay a supply line.

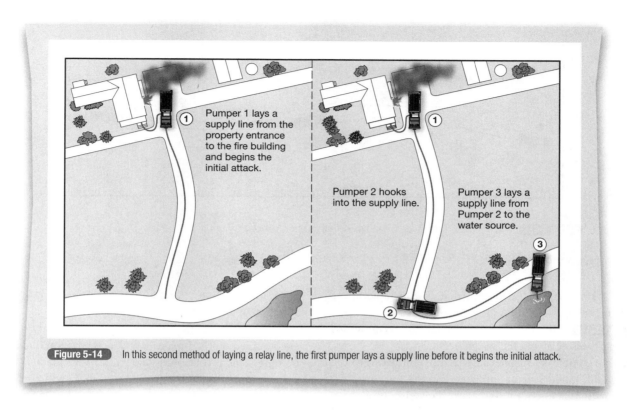

Figure 5-14   In this second method of laying a relay line, the first pumper lays a supply line before it begins the initial attack.

If an LDH is not being used for the initial supply line and if conditions permit, two supply lines should be laid initially to make sure the second supply line will be available if it is needed. Two supply lines should be laid initially only if they can be laid simultaneously.

Relay Pumping

After the relay pumpers are all in position and the supply lines are hooked up, pumping should begin. Supply lines should be charged from the source to the fire. The pump operator at the water source first charges the supply line to the next pumper. When the water reaches this second pumper, the operator opens the intake gate to charge the pump. The operator then opens the discharge to build up pressure to the next pumper. This pumper-by-pumper charging of the supply line continues until the pumper at the fire is receiving water.

After a pumper is discharging water, the pump operator must keep the discharge gates fully opened. The operator must also watch the pump discharge gauge, as discharge pressure will fall when the next pumper up the line begins taking water. As soon as it does, the operator should begin to adjust the throttle smoothly to bring the discharge pressure up to the starting pressure. The pump operator should continue to adjust the throttle until the pressure remains steady. This will keep fluctuations in pressures between pumpers to a minimum.

At the start of the relay operation, all pumpers in the line should be set to pump 150 psi to the pumper at the fire; that pumper should pump at the pressure required by the attack lines in use. The 150-psi pressure is easy to remember. It assures a good initial flow, and it will assure sufficient residual pressure to the pumper at the fire. Conversely, incoming pressure to the next pumper in the relay should never drop below 20 psi.

After the relay line is in operation, discharge pressures can be increased or decreased as necessary to meet firefighting requirements. The basic principle is to have all pumpers within the relay pump at the same discharge pressure. Good radio communications between the pumpers is essential for an efficient relay operation. Each pumper must be aware of the actions of the others so the operation can be coordinated.

Increasing Water Flow

If additional water is needed, another supply line should be laid between the operating pumpers. A separate, second relay

Key Points

After the relay pumpers are all in position and the supply lines are hooked up, pumping should begin.

Key Points

At the start of the relay operation, all pumpers in the line should be set to pump 150 psi to the pumper at the fire; that pumper should pump at the pressure required by the attack lines in use.

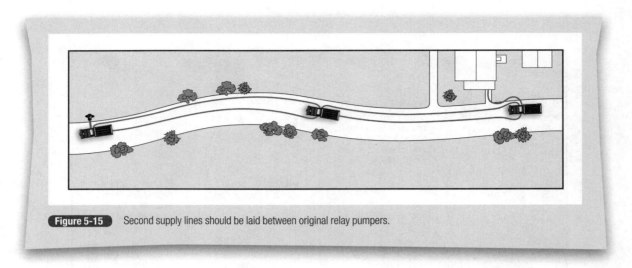

Figure 5-15   Second supply lines should be laid between original relay pumpers.

operation is not recommended immediately. It is much more efficient to build up the flow of water to the original attack lines **Figure 5-15**. After this is accomplished, and if conditions warrant, a second relay operation can begin.

Another option is to have these pumpers proceed to the fire to use tank water. Several pumpers or mobile water supply apparatus with full tanks represent a fairly large supply of water. There may be enough to allow additional attack lines to be used on the fire. This is especially important if hose lines are needed

in locations that are difficult to reach from the pumper being supplied by the relay.

Obviously, many different types of situations may be encountered on the fire ground. A combination of department training, standard operating guidelines, the incident management system, and a trained water supply officer will dictate which course of action the incident commander will use to successfully mitigate the incident.

Chief Concepts

- It is the task of the engine company to set up a continuous, uninterrupted supply of water at the fire ground.
- To be able to do so quickly and efficiently, the crew must be familiar with the sources of water that constitute an adequate and reliable water supply.
- This is a water supply that is sufficient every day of the year to control and extinguish anticipated fires in the municipality, at particular buildings, or building groups served by the water supply.
- "Being familiar" includes knowing the locations of hydrants, especially in areas of snowy winter weather, the location of static water systems, and/or other alternative sources of water.
- Capacities, volumes, flow rates, and the distances of these sources from structures should also be established.
- The engine company crew must also be able to make full use of their pumper to deliver water from the source to the fire ground.
- They should be aware of its true pumping capacity, in combination with whatever water sources are used, as well as its mechanical and operational limitations.
- The company must be aware of the water delivery capability of supply hose and how this capability changes with the size of the hose and the length of hose laid.

- An important consideration is the number of supply lines to be laid, which depends on the extent of the fire, the water source, the capacity of the pumper, the hose size, and the length of lay.
- Where excessively long hose lays require the use of pumper relays, the crew should be able to begin a relay and to take any position in a relay line.
- Every member of an engine company, whether fully employed, or on a call or volunteer fire department in a rural, suburban, or urban community, must be trained and thoroughly knowledgeable in the procedures for initiating and sustaining a viable water supply at a fire incident.

Key Terms

Pumper: Fire apparatus with a permanently mounted fire pump of at least 750-gpm (3,000 L/min) capacity, water tank, and hose body whose primary purpose is to combat structural and associated fires.

Residual pressure: The pressure remaining in a water distribution system while the water is flowing. The residual pressure indicates how much more water is potentially available.

Static pressure: The pressure in the water pipe when no water is flowing.

1. Water movement at fires is affected by
   a. Hose
   b. Pumper
   c. Water source
   d. All of the above affect the movement of water at fires

2. Flow test of the water main system and components should
   a. Be done by the fire department or the fire department should be part of the process
   b. Be done by the water department and the results made available to the fire department
   c. Be done by the property owner at private properties, with the test results shared with the fire department
   d. All of the above

3. Areas with _____ mains are particularly vulnerable to water supply problems.
   a. Gridded
   b. Dead end
   c. Looped
   d. Multisource

4. The pressure reading with the hydrant open and no water flowing through the pump is the _____ pressure
   a. Maximum or high
   b. Minimum or low
   c. Static
   d. Residual

5. The compound gauge is used to measure _____ pressure.
   a. Nozzle
   b. Discharge
   c. Static and residual
   d. All of the above

6. In regard to supplying fire attack hose lines from the apparatus water tank
   a. This is an acceptable water supply for small buildings, including most single-family detached dwellings.
   b. This is an acceptable water supply for single-family detached dwellings.
   c. The apparatus water tank should not be used to supply hose lines at structure fires.
   d. In smoke or fire showing situations, it is acceptable to begin the operation with the apparatus water tank as a supply, but an uninterrupted supply of water must be established.

7. When operating a tanker shuttle as a water supply
   a. Tankers should fill their tanks using portable pumps.
   b. Tankers should take suction to fill their tanks.
   c. Pumpers should be stationed at the water source to fill tankers.
   d. Any of the above could be equally effective depending on the water source and access to the static source of water.

8. The minimum pump capacity for a pumper is _____ gpm.
   a. 500
   b. 750
   c. 1,000
   d. 1,250

9. A pumper must be capable of providing 100% of its rated capacity at _____ psi pump pressure when at draft.
   a. 100
   b. 150
   c. 200
   d. 250

10. Pumper intakes in the United States must have
    a. Male threads
    b. Female threads
    c. Either male or female national standard threads
    d. Either male or female threads used by the jurisdiction where the pumper is located

11. When increasing the engine speed does not result in an increase in discharge pressure, the apparatus fire pump is
    a. Operating at peak efficiency
    b. Operating at peak capacity
    c. Cavitating
    d. Defective

12. The residual pressure from a hydrant should not be allowed to drop below _____ psi, except under extreme circumstances.
    a. 1
    b. 5
    c. 10
    d. 20

13. When pumping into an LDH, a pressure relief device should be provided with a maximum setting of _____ psi.
    a. 150
    b. 200
    c. 250
    d. 275

14. The standard fire hose for the fire service is the _____ diameter line.
- **a.** 1¾ inch
- **b.** 2 inches
- **c.** 2½ inches
- **d.** There is no standard fire hose size.

15. Generally, the maximum flow from a hand-held attack hose is 350 gpm. At this flow, the friction loss is approximately _____ psi per 100 foot of 2½-inch hose.
- **a.** 5
- **b.** 10
- **c.** 20
- **d.** 25

16. A pumper is being supplied with dual 2½-inch hose lines. A single 3-inch supply would be _____ in water-carrying capacity.
- **a.** Approximately equal
- **b.** Slightly higher
- **c.** Approximately double
- **d.** Approximately triple

17. An LDH can be used
- **a.** As a water supply
- **b.** As an attack line
- **c.** To supply attack lines
- **d.** Both as a water supply and to supply attack lines

18. A single 5-inch supply hose is approximately equal to _____ 2½-inch hoses.
- **a.** two
- **b.** four
- **c.** six
- **d.** eight

19. The proper size hose for a supply line
- **a.** Depends on the actual flow needed, but 2½-inch hoses (single or dual lines) are usually sufficient
- **b.** Is 2½ or 3 inches for single-family detached dwellings, but nothing less than 3 inches for larger buildings
- **c.** Should be a 3-inch hose or larger
- **d.** Should always be an LDH if available

20. When pumping from one pumper to another to relay water from a water supply to the pumper at the fire location, pumpers in the relay except the pumper at the fire should have a pump discharge pressure of _____ psi.
- **a.** 100
- **b.** 150
- **c.** 200
- **d.** 250

21. When pumping from one pumper to another to relay water from a water supply to the pumper at the fire location, pumpers in the relay except the pumper at the fire should have a pump intake pressure no less than _____ psi.
- **a.** 5
- **b.** 10
- **c.** 15
- **d.** 20

Chapter 6

Learning Objectives

- Analyze the concept of using a direct attack during offensive fire attack scenarios.

- Analyze the advantages and disadvantages of using an indirect attack during offensive fire scenarios.

- Analyze the concept of using a combination attack during offensive fire scenarios.

- Examine factors to determine the choice of the initial attack lines and nozzles used on the fire ground.

- Recognize the various types of scenarios that will assist in selecting the proper attack line and advancing it to a proper location within the building.

- Recognize the conditions that indicate the presence of a smoldering fire, a condition that could be disastrous if not handled properly.

At a working structure fire, the placement and operation of initial attack lines protect occupants and fire fighters as well as providing the first water on the fire for extinguishment. The incident commander (IC) should conduct a risk versus benefit analysis before the decision is made to allow fire fighters to enter a building to conduct an aggressive interior attack. If a defensive mode of operation is used, fire fighters must take up safe positions to protect exposures and operate on the fire building from the outside. Fire fighters must make a decision as to the type of initial attack line that will best manage the situation at hand. Initially choosing the correct hose line will protect fire fighters and victims and enable crews to control the fire. Choosing smaller, ineffective hose lines for initial attack usually will not control the fire and create negative results until the proper hose lines are used. This potentially places everyone on the fire ground in a dangerous situation.

Most fire departments operate pumpers with preconnected hose lines with designated hose sizes, lengths of hose lines, and specific types of nozzles. These hose lines are suitable for initial attack lines for the majority of the times they are used. Unfortunately, many fire departments have become complacent and stretch these hose lines at every incident. A company officer must be able to choose the correct hose line and nozzle for the intended application. This may mean deviating from the norm and operating a hose line that will provide safety to fire fighters and victims while controlling and extinguishing the fire.

Fire fighters must operate within an incident management system and coordinate efforts between engine and ladder company personnel. The fire ground must be managed so that engine company personnel are prepared to enter the building when it has been ventilated. Ventilation must be coordinated with suppression efforts and a coordinated fire attack will ensure that both tasks take place simultaneously.

With pumpers well positioned and an adequate water supply assured, the outcome at the fire ground is dependent, to a great extent, on the effectiveness of the initial attack. The success of the initial attack most often centers on the effectiveness of the company officers' decisions regarding the size, number, and placement of attack lines, as well as the types of nozzles to be used against the fire.

The most common type of fire encountered by responding fire fighters is the free-burning fire. The fire is free burning when all three requirements for combustion—oxygen, fuel, and heat—are present in sufficient quantities to promote the spread of flames and fire.

An interior attack is an offensive operation in which fire fighters enter a building with an attack line to control and extinguish the fire. If there is any chance that a fire building is occupied and an interior attack is possible, the fire should be fought from within. Fire fighters should get inside to attack the seat of the fire, place attack lines between the victims and the fire, conduct a primary search for victims, and ventilate promptly.

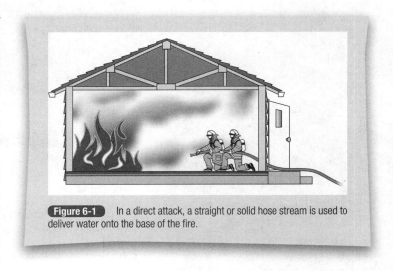

Figure 6-1   In a direct attack, a straight or solid hose stream is used to deliver water onto the base of the fire.

An interior attack can be conducted using a direct, indirect, or combination method.

Direct Attack

In almost all offensive fire attack situations, a direct attack is used. A direct attack is the surest way of controlling the fire and minimizing the danger to occupants **Figure 6-1**. A **direct attack** is one in which a solid or straight hose stream is used to deliver water directly onto the base of a fire. This cools the burning material below its ignition temperature and extinguishes the fire. A direct attack works best if the fire attack is coordinated with engine companies advancing and operating attack lines and truck companies performing ventilation.

Indirect Attack

A free-burning fire may also be attacked indirectly. An **indirect attack** is one in which a solid, straight, or narrow fog stream is used to direct water at the ceiling to cool superheated gases in the upper levels of the room **Figure 6-2**. The objective of the indirect attack is to prevent flashover by removing heat from the upper atmosphere. This method injects a stream of water into the heated upper levels of the room. The water is converted to steam, which absorbs a tremendous amount of heat. This process quickly drops the temperature in the atmosphere.

The disadvantage of an indirect attack is that the water-to-steam expansion ratio is capable of causing serious burns to fire fighters and victims. Because of this, an indirect attack is a poor choice in areas where victims may be located. It will certainly make rescue attempts more difficult, or impossible, after this method is applied. Using this attack method may be better suited for unoccupied areas such as basements, attics, or storage areas. Fire fighters should shut down the attack line as soon as enough water has been discharged to cool the area. If too much water is converted to steam, it will push the steam and hot gases down

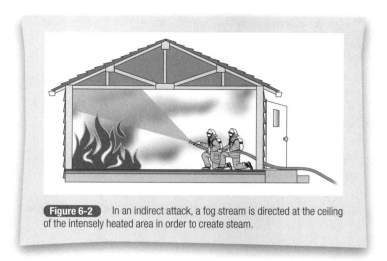

Figure 6-2 In an indirect attack, a fog stream is directed at the ceiling of the intensely heated area in order to create steam.

toward the floor where they are operating. When the area has been cooled down and ventilation has taken place, the operation should be changed from an indirect attack to a direct attack.

Combination Attack

The **combination attack** uses both the direct and indirect methods one after the other. If the room is extremely hot and nearing flashover conditions, the indirect attack is used to bring down the temperature. After this has been accomplished, the direct attack is used to extinguish the main body of fire. A limited amount of water should be used during the indirect attack to limit the amount of steam produced, which may force fire fighters out of the area. Before fire fighters enter the building to conduct firefighting tasks, a risk versus benefit analysis should be conducted.

Sometimes responding companies will find that a fire is smoldering rather than free burning. The exact position of the fire will not be evident, as no flames will be showing. The sequence of operations to be used against a smoldering fire is different from that used against a free-burning fire. The differences are covered in the last section of this chapter.

Keep in mind that an aggressive interior attack should not be made in buildings that are in various stages of demolition, have been abandoned for long periods, have been burning for a predetermined period of time, contain construction features detrimental to safe interior operations, have had previous fires, or are under construction. The IC should conduct a risk versus benefits analysis before the decision is made to allow fire fighters into a building to conduct firefighting activities, including a primary search.

If any interior operations are to be made in these buildings, they should be carried out only after the fire has been knocked down from the outside and a careful check has been made on the condition and relative safety of the structure on the inside.

The normal aggressive interior attack should be made on those buildings that are in use and especially those that are occupied at the time of the fire. Even in this case, however, if a large intense fire is encountered, it may be necessary to knock down or control the fire from the outside using solid or straight streams before an interior attack can be made.

Remember that occupants will benefit most by the extinguishment of the fire. If the fire is of such intensity that an interior attack cannot control or extinguish the fire with a sufficient amount of attack lines that are appropriately sized, then a defensive attack should be conducted from the outside. If an interior attack is to be made on a free-burning fire, several other decisions will need to be made.

Choosing Attack Lines

One size of hose or one type of stream is certainly not the answer to every fire situation that confronts an engine company on arrival at the fire ground. The choice of initial attack lines and nozzles depends on the purpose of the attack, whether it be a holding action, exposure protection, a defensive operation, or an offensive operation with an interior attack on the main body of the fire.

Factors that affect this choice include the size and location of the fire, how the attack lines are to be used against the fire, available equipment, and the personnel available for fire attack. Based on these factors, the following decisions must be made:

- What size hose diameter is appropriate?
- What type of nozzle should be used?
- How many hose lines are needed?
- Where will hose lines be positioned?
- In what type of operation, offensive or defensive, will the hose lines be used?

Sizes of Attack Lines

Choosing the size or sizes of attack lines depends on the extent and location of the fire and how it will be fought. Most fire departments use 1¾- and 2½-inch hose as attack lines for an interior attack.

The 1¾-inch hose line is easy to maneuver and to advance with limited personnel. After the attack line is in position and charged to the proper nozzle pressure, one fire fighter generally can handle the nozzle. (It is dangerous, however, to assign only one person to an attack line for interior firefighting, and this practice should not be allowed.) The typical 1½-inch attack line is equipped with a spray nozzle that will discharge, under proper pressure, from 60 to 125 gpm.

The mobility of the 1½-inch attack line has made it popular among fire fighters. It can be advanced quickly and easily to the fire area and can be very effective against small fires; however, there is a tendency to use it against larger fires, where it is much less efficient or a complete failure, wasting water, time, and effort, and causing avoidable injuries to fire fighters as well as burning down the building. This tendency should be avoided. Because it is as easy to place a 1¾-inch handline in service as compared with a 1½-inch handline, which flows considerably less water, it is recommended that the use of a 1½-inch hose be eliminated for structural firefighting. The theory is simple: If you have a big fire, you need big hose lines.

The 1¾-inch attack lines, which are as mobile and as easy to handle as the 1½-inch line, can discharge 120–200 gpm with either a smooth-bore nozzle or a spray nozzle. This permits the attack line to be used safely and efficiently on somewhat larger fires than the 1½-inch attack line. The 1¾-inch hose line has gained wide acceptance in the fire service as an initial attack line.

Here again, however, fire fighters should avoid the tendency to use a 1¾-inch hose against larger fires where it may not be effective. Fire fighters need to stop pulling one type of attack line off the truck for every fire situation. If a fire cannot be controlled or extinguished with the hose lines that are being operated, then additional, larger hose lines will be needed.

Because the 2½-inch attack line is heavier than the 1¾-inch attack line, it is not as easily manipulated. Before the line is charged,

however, it, too, can be readily advanced and maneuvered; practice in handling the hose line overcomes difficulties in using it. Adequate staffing levels will get the hose line into service more quickly.

Although more effort is required to get a 2½-inch attack line into position, the payoff, in terms of water delivery for fire control, is worth the effort. This is especially true if the fire has advanced beyond the penetrating capability of a 1¾-inch attack line. The 2½-inch attack line, when equipped with a spray nozzle operating at a nozzle pressure of 100 psi or a solid-bore nozzle with a 1⅛-inch tip operating at a nozzle pressure of 50 psi, will discharge at least 250 gpm when operated properly. Its much larger stream absorbs more heat and reaches further, providing better fire control in less time.

Attack Lines and Nozzles

Smooth-bore nozzles producing solid streams have greater reach and more water delivery capability than spray nozzles producing straight streams. The nozzle most often used on 2½-inch attack lines is equipped with the 1⅛-inch smooth-bore tip with a 265 gpm discharge rate at a nozzle pressure of 50 psi and the 1¼-inch smooth-bore tip with a 325 gpm discharge rate at 50 psi. These solid stream tips require only half the nozzle pressure of the 2½-inch spray nozzle, in spite of their greater reach and water delivery; however, they are more difficult to handle than the spray nozzle owing to the increased water flow. Again, practice with these hose lines and nozzles will increase operating efficiency and overcome any reluctance to use them (**Figure 6-3** shows a comparison of nozzles and streams).

Solid Streams Versus Straight Streams

For the safest and most effective operations for interior firefighting where civilians or fire fighters will be in the area, smooth-bore nozzles producing a solid stream or spray nozzles adjusted to the straight stream pattern should be used. This will aid rescue and fire control because these stream patterns are less likely to disturb the thermal layer than if a fog pattern was used. This may help with visibility as well as preventing water from turning to steam, which could scald both fire fighters and fire victims.

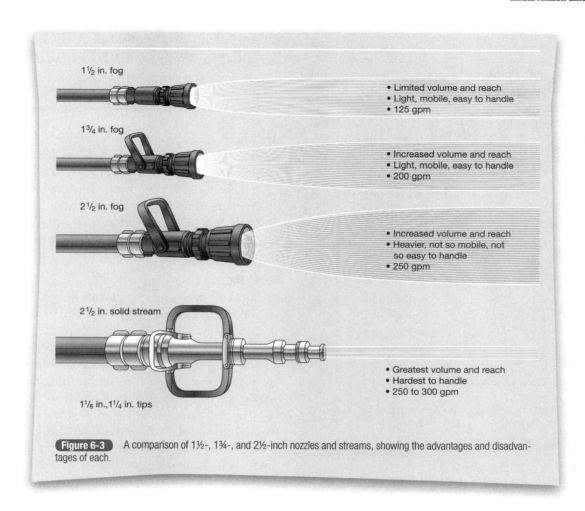

1½ in. fog
- Limited volume and reach
- Light, mobile, easy to handle
- 125 gpm

1¾ in. fog
- Increased volume and reach
- Light, mobile, easy to handle
- 200 gpm

2½ in. fog
- Increased volume and reach
- Heavier, not so mobile, not so easy to handle
- 250 gpm

2½ in. solid stream
- Greatest volume and reach
- Hardest to handle
- 250 to 300 gpm

1⅛ in., 1¼ in. tips

Figure 6-3 A comparison of 1½-, 1¾-, and 2½-inch nozzles and streams, showing the advantages and disadvantages of each.

The use of spray nozzles adjusted on a fog pattern inside a building should be restricted to unoccupied confined spaces such as unoccupied basements, attics, or storage areas. Here, with heavy fire involvement, a scuttle can be removed, a stairway pulled down, or a hole opened in the ceiling below with a pike pole. The fog stream can be placed into the attic space to let the cooling and steam conversion do their work. This should be followed by venting the area and checking for spot fires.

When it is known that the building is adequately ventilated opposite the direction from which the spray nozzle is being advanced, a fog stream can be used. In this case, however, it should be operated at no more than the 30° angle, not at a wider setting. The 30° angle produces a combination of reach and fog pattern. Increasing the angle to produce a wider pattern gives the fire fighter more protection but decreases the reach of the stream, diminishes firefighting effectiveness and pushes the fire through the building. Solid streams and straight streams move little air compared with a fog stream. A wide-angle fog stream does not penetrate a fire. In fact, it can have the detrimental effect of pushing fire, heated gases, and smoke into uninvolved areas of the building **Figure 6-4**.

Many fire fighters, however, have found that even the 30° pattern cannot control a hot interior fire and that the simple action of adjusting the angle to a straight stream can result in better reach and penetration.

A fog stream may be used during an indirect attack to absorb high levels of heat. When this is accomplished, a direct attack should then be used. Otherwise, a fog stream should not be

Key Points

Smooth-bore nozzles producing a solid stream or spray nozzles adjusted to the straight stream pattern are the safest and most effective for interior firefighting where civilians or fire fighters will be in the area.

Key Points

Spray nozzles adjusted on a fog pattern inside a building should be restricted to unoccupied confined spaces such as unoccupied basements, attics, or storage areas.

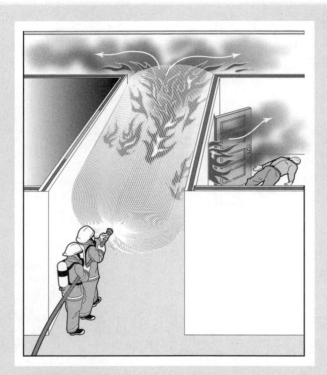

Figure 6-4 A wide-angle fog stream can sometimes push fire, heated gases, and smoke into uninvolved areas of the building.

used unless there is adequate ventilation. If a fog stream is used incorrectly, there is a possibility that the fire, heated gases, and smoke as well as steam can come back onto the fire fighters operating the attack line.

Smooth-bore nozzles should be used where solid streams are necessary to control the fire. Remember that a straight stream from a spray nozzle is still a broken stream, a stream of water separated into coarsely divided droplets.

The controversy over solid streams versus fog streams has raged for many years. It is not the intention here to resolve this argument but only to point out the differences between the two stream patterns.

Unfortunately, in years past, too many fire departments have abandoned the use of smooth-bore nozzles and 2½-inch hose for attack line operations. This is a big miscalculation. Solid streams provide reach, penetration, and lower nozzle pressure and should be considered for both offensive and defensive operations. Fire

Key Points

Smooth-bore nozzles should be used where solid streams are necessary to control the fire. A straight stream from a spray nozzle is still a broken stream.

Key Points

In almost all cases, spray nozzles are designed to deliver their rated flow capacity at 100 psi.

departments using only 1¾-inch hose for attack lines are at a disadvantage when confronted with a large fire and being unable to apply large quantities of water through a hose line.

On the 1¾-inch attack lines, the smooth-bore ^{15}/16-inch tip (180 gpm at 50 psi nozzle pressure) or the 1-inch tip (flowing 200 gpm at 50 psi nozzle pressure) is being used with much success over comparable spray nozzles, which operated at a higher nozzle pressure.

Spray Nozzles

In almost all cases, spray nozzles are designed to deliver their rated flow capacity at 100 psi. Currently, "low-pressure" spray nozzles are available that operate at 75 psi nozzle pressure. Four types of spray nozzles are commonly available and in use within the fire service. These are the basic spray nozzle, constant gallonage spray nozzle, constant pressure (automatic) spray nozzle, and the constant/select gallonage spray nozzle.

The basic spray nozzle is an adjustable-pattern spray nozzle in which the rated discharge is delivered at a designed nozzle pressure and nozzle setting. The constant gallonage spray nozzle is an adjustable-pattern spray nozzle that discharges a constant discharge rate throughout the range of patterns from a straight stream to a wide spray at a designed nozzle pressure. The constant pressure (automatic) spray nozzle is an adjustable-pattern spray nozzle in which the pressure remains relatively constant through a range of discharge rates. This nozzle is designed to maintain 100 psi at the tip regardless of the flow. For instance, if 200 gpm is flowing at 100 psi and the flow is reduced to, for example, 125 gpm, the nozzle will adjust to produce 100 psi nozzle pressure and maintain a "good-looking stream."

Fire fighters must realize, however, that this stream might be ineffective on the fire. It is the gpm flow that controls a fire, not the nozzle pressure; therefore, if 200 gpm is required to control the fire, the 125 gpm "good-looking stream" will not do the job. With the basic spray nozzle and the constant gallonage spray nozzle, a weak stream is readily apparent.

The constant/select gallonage spray nozzle is a constant discharge rate spray nozzle with a feature that allows manual adjustment of the orifice to effect a predetermined discharge while the nozzle is flowing.

Effective Stream Operation

Some guidelines for safe and efficient stream operation are listed here:

- Use smooth-bore nozzles on 1¾- and 2½-inch preconnected attack lines or adjust spray nozzles on the straight stream pattern.

- Conduct a risk versus benefit analysis before entering a building.
- Crack the nozzle to bleed the air out of the hose line ahead of the water.
- Before the door to a fire area is opened, all fire fighters at that location should be positioned on the same side of the entrance and remain low.
- Use a direct attack to deliver water directly onto the fire.
- Stay low when entering the fire area to let the heat and gases vent before moving in.
- If fire shows at the top of the door as it is opened, the ceiling should be hit with a solid stream to cool it and control fire gases.
- Do not open the nozzle until fire can be seen unless fire fighter safety is involved. Water directed at smoke will make visibility and smoke conditions worse; a fog stream will greatly increase this problem. Remember that fire fighters are much more tolerant of dry heat than they are of wet heat.
- Direct the stream at the base of the fire if it is localized. If the area is heavily involved, direct the solid stream forward and upward at the ceiling in a side-to-side motion or rotate the stream.
- As the advance is made, the angle of the stream should be lowered and an attempt made to attack the main body of fire.
- Sweep the floor with the stream to cool burning debris and hot surfaces. This will help to prevent burns to hands, knees, PPE, and the attack line as it is advanced.
- When the main body of fire is knocked down, shut down the stream, and let the area "blow" (allow smoke and gases to rise and vent). Listen for crackling sounds, and look for areas that "light up" so that extinguishment can be completed.
- When the fire is knocked down, shut down the attack line. This will help control water damage and the weight of the water on the floor.

After entering an area that is very hot and finding no fire, check the area below, adjacent areas, walls, and ceilings and other vertical and horizontal voids for fire extension. The IC should be notified of the conditions within this area. It may be advantageous to move the fire fighters out of this area if the fire can't be found. From an adjacent location, an attack line could protect the building or means of egress from the building.

Key Points

Three sizes of hose lines allow a variety of tactics in fire attack. The 1½- and 1¾-inch attack lines are mobile with limited water delivery capacity and reach; the 2½-inch line is heavier and not as easy to move or handle when it is charged, but it delivers greater water capacity and reach.

Number of Lines

The three sizes of hose lines allow a variety of tactics in fire attack. The 1½-inch and 1¾-inch attack lines are mobile with limited water delivery capacity and reach; the 2½-inch attack line is heavier and not as easy to move or handle when it is charged, but it delivers greater water capacity and reach. Which attack lines are used and how they are operated will depend on the fire situation, including the size and location of the fire. As noted previously, a fire department using a 1½-inch hose should consider purchasing a 1¾-inch hose as a replacement. The carrying capacity can be doubled using a 1¾-inch hose with little difference in weight and ease of operation.

Hose lines will be needed to attack the main body of the fire and stretched to areas above and below the fire as well as into adjoining areas of the building. Hose lines also may be needed to protect exposure buildings. If 1¾-inch lines can accomplish these tasks, they should be used (**Figure 6-5** and **Figure 6-6**). If any doubt exists about the effectiveness of these smaller attack lines, a 2½-inch attack line should be placed in service immediately. Often a combination of the two different attack lines works well. For example, a 1¾-inch attack line can be used for the initial attack on the main body of fire. This attack line can be backed up by a 2½-inch hose line, whereas a 1¾-inch attack line is taken to the floor above the fire.

Also, see Chapter 5 for more information on hose line sizes. As another example, in a building with several floors, 1¾-inch hose lines can be ideal for getting above the fire to stop vertical spread from floor to floor by way of the windows, whereas 1¾ inch or 2½ inch can be used to attack the main body of fire.

If there was a large fire in an open area inside a department store or supermarket and an offensive operation was underway, the main body of fire would be attacked with 2½-inch attack lines; 1¾-inch attack lines may be useful in keeping the fire from spreading horizontally through doors or other openings into unaffected areas. These smaller attack lines would be backed up by additional 2½-inch hose lines.

Master Streams for Initial Fire Attack

Occasionally, a fire will be so large that master stream appliances must be used for initial attack. There would be little sense in trying to control such a fire with hand-held attack lines. If an exterior attack is warranted, master streams should be placed in service as soon as possible. During this defensive mode of operation, offensive operations must be curtailed, and fire fighters should not be allowed to enter the structure.

If an exterior attack is warranted on a large fire and hand-held attack lines cannot control it, then heavy streams should be placed in service as soon as possible. Master streams are the safer and more effective option for a defensive attack. As recommended in Chapter 2, a master stream appliance mounted on a pumper or portable monitor operated from the ground should be placed in

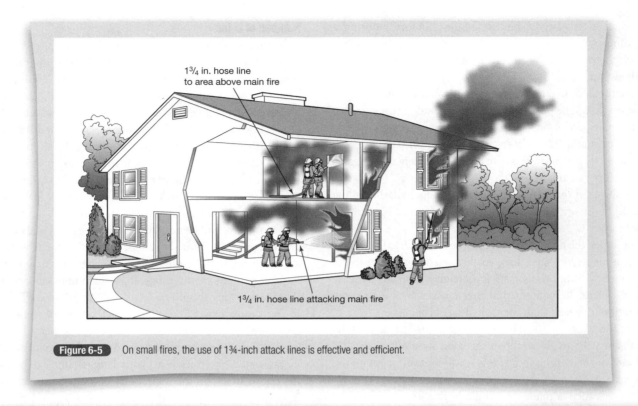

Figure 6-5   On small fires, the use of 1¾-inch attack lines is effective and efficient.

Figure 6-6   A combination of 1¾- and 2½-inch attack lines is usually efficient and effective in fighting a larger fire.

Figure 6-7   When a large fire is encountered, a prepiped or preconnected master stream appliance can be placed in service quickly to attack the fire.

service. This appliance could be initially operated with tank water as soon as it was in position. A continuous water supply should then be established as soon as possible. Master stream appliances can also be operated from aerial fire apparatus **Figure 6-7** .

Initial Attack Operations

Older parts of almost every community have combination commercial and apartment buildings. These structures have stores on the first floor and one or more floors of apartments above them. A fire in a store can create serious fire and rescue problems in the floors above. The following are typical scenarios.

Attack Lines

Fire fighters arrive to find fire in one of the stores. The IC realizes that it is imperative to control the main body of fire as quickly as possible. If entry can be made and the store is small, 1¾-inch attack lines might be adequate for controlling the fire. If the store is well involved (perhaps fire is extending out of the front of the store), however, a 2½-inch attack line should be used on the store fire, and 1¾-inch attack lines should be advanced immediately above the fire to protect occupants of the apartments, assist in rescue, and cut off any vertical fire spread. With 50% involvement of a commercial space, use 2½-inches as your interior attack line.

Attacking the well-involved store fire with 1¾-inch hose lines would be a waste of time and effort in a situation in which both are at a premium. The heat and force of the extending fire would turn away smaller streams, allowing little penetration into the fire resulting in insufficient fire control **Figure 6-8** . The 1¾-inch attack lines should be shut down, and a sufficient number of 2½-inch attack lines should be placed in operation on the main body of fire.

The 2½-inch solid stream will penetrate into the heated area and, with its greater reach and volume, will have a better chance of controlling the fire. A 2½-inch preconnected attack line should not take much more time to be placed into operation than a smaller attack line with a sufficient amount of fire fighters present.

Fires in the lower floors of an occupied structure must be knocked down as quickly as possible **Figure 6-9** . Engine crews assigned to controlling the main body of fire should not be assigned to other areas of the building until this task is accomplished. Occupants of the building will benefit as much from the control of the main body of fire as from any other action.

In modern garden apartments, fires are often encountered in basement storage rooms. These are usually large open areas divided into small storage cubicles for individual tenants. The same problem exists here as in the storefront apartment house: a fire under occupants in apartments above.

If the fire in the storage area has gained considerable headway, it should be attacked with 1¾- or 2½-inch hose lines and a 2½-inch backup line. Fire fighters should be assigned to the floors above to conduct the primary search and perform rescue operations if needed. The 1¾-inch attack lines should be advanced to these areas to stop any vertical fire spread and to protect searching fire fighters and victims.

Because it may be difficult to thoroughly ventilate a basement, solid or straight streams from 1¾- or 2½-inch attack lines should be used. This is especially true in those basements built like concrete vaults, with concrete walls, floors, and ceilings with no means to quickly vent to the outside. In such cases, the safest and most efficient attack is with 2½-inch smooth-bore nozzles from an outside doorway, bulkhead, or other opening. This stream allows greater reach into the area, provides heavy water flow into the fire, and keeps the fire fighter from being subjected to high concentrations of heated gases and steam while accomplishing fire control. If an interior attack is made, fire fighters must protect the opening and the stairway, if one is

Key Points

Attacking a well-involved store fire with 1¾-inch lines would be a waste of time and effort. The 2½-inch solid stream will penetrate into the heated area and, with its greater reach and volume, will have a better chance of controlling the fire.

(a) Fire attacked with a 1¾ in. attack line   (b) Fire attacked with a 2½ in. attack line

Figure 6-8 A 1¾-inch attack line (a) cannot penetrate a large, hot fire, but a 2½-inch attack line (b) is capable of gaining control.

located nearby. The basement may be difficult to enter unless ventilation is accomplished. Fire fighters must be well supervised and be aware of the danger that this situation presents. Fires in similarly constructed areas other than basements should be fought in the same way.

The general attack procedures given in these two examples can be used in almost any kind of structure, with the fire in

Figure 6-9 An engine crew controls the main body of fire below grade.

any location. For instance, where fire has gained control of a good portion of a large floor area, 2½-inch attack lines should be used on that floor. The 1¾-inch attack lines should be used on the floors above, around, and under the fire, to prevent fire extension and, if necessary, to assist in rescue and evacuation. This usually results in the most efficient firefighting operation accomplished with the fewest personnel. The important point is to use the proper sizes of hose and types of nozzles to confine and extinguish the fire at all points within the building. As soon as it becomes obvious that a hose stream is not accomplishing its objective, get a bigger one. If more are needed, get them.

Advancing Attack Lines

In the previous examples, fires were on the first floor or in the basement of a particular building; however, the fire could have started on any floor of a multistoried building, and attack lines must be advanced up to and above the fire floor. Moreover, even if the fire is on the first floor, hose lines, must be advanced to the floors above the fire to protect the primary search and rescue efforts as well as to confine and extinguish the vertical spread of

Key Points

The proper size hose and nozzle must be used to create a stream of water to confine and extinguish the fire at all locations within a building.

Hose lines must be advanced to upper floors of a fire building.

the fire. Whatever reason, hose lines must be advanced to upper floors of the fire building.

On fires above the first floor, the first-arriving engine companies usually will make use of stairways for getting attack lines up to and above the fire, as stairways give them quick access to the building. As they advance their attack lines, the fire fighters must extinguish any fire that has moved into a stairway. Control of stairways is important—both to eliminate them as channels for the vertical spread of fire and to keep them open as escape routes for building occupants and fire fighters **Figure 6-10**. If there are multiple structures, use one for fire attack and one for evaluation.

Additional arriving companies can also use the stairways if they are safe and wide enough; however, the stairs may be narrow or clogged with hose lines or poorly located in relationship to the position that a company must take. Other means of advancing hose lines to upper floors are available and should be used when necessary. Hose lines may be advanced using the following methods:

- By means of ground ladders, aerial ladders, or aerial platforms
- By hoisting with ropes

Figure 6-10 An engine crew controls the main body of fire and protects the stairway, whereas a ladder crew performs search and rescue operations.

Control of stairways is important—both to eliminate them as channels for the vertical spread of fire and to keep them open as escape routes for building occupants and fire fighters.

Stairways may not be the best means of advancing hose lines to upper floors. Many other methods are available and should be used when necessary.

- By being carried into the building and then connected to an outside hose line through a window, balcony, or porch
- By being passed up to a window with a pike pole or shepherd hook

Too often, such methods are detailed in training manuals, practiced during training sessions, and then forgotten on the fire ground.

These alternative methods may require much less hose than if advancing up stairways. When hose lines are taken up the face of the building, only one section of hose is needed for every four or five stories. In a stairway, one 50-foot section must be allotted for each floor because so much length is taken up in winding around stairs and through hallways.

Ventilation

All of the operations discussed in this section are enhanced by proper ventilation, which allows fire fighters on the hose lines to move more easily and quickly to the proper firefighting positions. Ventilation of stairways and hallways is of utmost importance. Venting the stairs can result in unintended consequences, as fire and smoke can be pulled into the stairway from fire on lower floors. Engine company members should position attack lines in key positions to protect fire fighters advancing to upper floors of the building. Fire fighters attempting to advance attack and backup lines to the fire should vent as they go if that action will help them to advance. For a free-burning fire, the general procedure is to use a coordinated fire attack, which will provide ventilation and fire attack simultaneously. **Figure 6-11** shows a coordinated fire attack in which one team of fire fighters attacks the fire directly while another team advances to the roof to ventilate the area above the fire.

Ventilation of stairways and hallways is of utmost importance. Fire fighters attempting to advance lines to the fire should vent as they go if that action will help them to advance.

Figure 6-11 This coordinated fire attack includes ventilation and extinguishment.

Smoldering Fires

Fire fighters responding to an alarm may encounter a fire that is not free burning or at least does not appear to be. There will be enough smoke to indicate that there is a fire, but there will be no flames. In such a case and lacking information about the exact situation, the fire fighters must assume that they have come upon a smoldering fire. If it is not handled properly, this type of fire can be disastrous.

Indications of a Smoldering Fire

A smoldering fire can be indicated by one or more of the following conditions:

Key Points

If a fire appears to not to be free burning, fire fighters must assume that it is a smoldering fire and handle it as such. If not handled properly, a smoldering fire can be disastrous.

- Smoke is visible, but little or no fire is visible from the outside.
- The smoke rises rapidly as it comes from the building, indicating that it is hot (however, humid weather may hold down the smoke).
- Usually the building will be tight with windows, doors, and other openings to the outside secured.
- Smoke leaves the building under pressure from around windows, doors, eaves, or other openings.
- The smoke may be yellow or dirty brown in color.
- Although no flames are showing, the window glass is brown or stained from heavy carbon deposits created by the smoke.
- There are signs of extreme heat present.
- All windows are darkened with linear cracks.
- Smoke will exit the building and appear to be sucked back into the building (i.e., breathing).

Whenever any of these conditions is present or appears to be present, the fire must be handled as a smoldering fire for the safety of fire fighters and for proper firefighting operations.

A smoldering fire has sufficient heat and fuel to become free burning. The heat comes from the fire, which was probably burning freely at one time. The fuel is mainly carbon monoxide gas from the original fire and the smoldering fire, along with the contents of the building and, perhaps, the building itself. The carbon monoxide has filled the building and surrounded the smoldering fire, cutting it off from an unlimited supply of oxygen. Thus, a smoldering fire needs only oxygen to burst into flame. A fire may be smoldering in a building of any size or type or, in some cases, in only one area of a large structure such as an attic, storage area, or other concealed areas.

Backdraft

A smoldering fire must be ventilated before it is attacked; that is, the carbon monoxide must be cleared from the building before air is allowed to enter it. The addition of any oxygen to the heat and fuel, even as little oxygen as might enter the building when an outside door is quickly opened and closed, will lead to immediate ignition of the fire.

The sudden ignition can take any of several forms. In one situation, the gases and preheated combustibles might simply burst into flames, engulfing the building in fire. In another, the force of the ignition might be enough to blow windows, doors, and fire fighters out of the building. There also could be an explosion strong enough to cause structural damage to the building and grave injuries to fire fighters and any other occupants. Just what will happen cannot be determined beforehand, but it is certain that the addition of

Figure 6-12 Backdraft occurs when oxygen is introduced into a superheated space previously deprived of oxygen.

Figure 6-13 Fire fighters should ventilate a smoldering fire at the highest point possible.

Key Points

Attack lines should be charged and ready for use during the ventilation process. Fire fighters on the attack lines should be in safe positions and be ready to enter the building as soon as ventilation is completed.

oxygen will cause some sort of ignition. That ignition is referred to as backdraft **Figure 6-12**. Further defined, a backdraft is the sudden explosive ignition of fire gases when oxygen is introduced into a superheated space previously deprived of oxygen.

Ventilation of a Smoldering Fire

The carbon monoxide, other heated gases, hot air, and smoke will have been convected upward and will be collecting at the top of the structure. An opening must be made on the building, at as high a point as is safely possible, to release these gases and allow them to move out of the fire area **Figure 6-13**. This will relieve the explosive situation and greatly reduce the chance of backdraft. Again, this ventilation must take place before any attempt is made to enter the building. Otherwise, air entering the building with fire fighters will cause immediate ignition. These tactics must be coordinated between engine companies preparing to enter the structure to attack the fire and ladder companies who must perform proper ventilation procedures before entry is made.

It is important to ventilate fully and in the right places to ensure that the hot gases are dispersed. On the one hand, ventilation should not be rushed or haphazard. On the other hand, it should be completed promptly. Heat is, and has been,

Key Points

A smoldering fire must be ventilated before it is attacked; that is, the carbon monoxide must be cleared from the building before air is allowed to enter it. The addition of any oxygen to the heat and fuel, even a minute amount, will lead to immediate ignition of the fire, or backdraft.

building up inside the structure. It is possible for the heat itself to break the window glass and allow air to enter before ventilation is completed. The result would again be sudden ignition; this time, however, fire fighters would be near or on the building.

Attack lines should be charged and ready for use during the ventilation process. Fire fighters on the attack lines should be in safe positions, protected from flying glass, and ready to enter the building as soon as ventilation is completed. Apparatus should not be placed in a direct line with the building.

Initial Attack

After the gas, hot air, and smoke are released from the building, outside air will enter and will cause the previously smoldering fire to burst into flame. This is a sign that venting is complete. The danger of explosion will be minimal, but fire fighters will have only a limited idea of the size of the fire.

At this point, initial attack should begin. Entering fire fighters will allow additional air to reach the fire, and it may be burning freely by the time initial attack is begun. Attack lines should be stretched above and around the fire, and search and rescue operations should be initiated.

Key Points

After the building is vented, outside air will cause a smoldering fire to burst into flame. As this point, the initial attack should begin.

Chief Concepts

- At a working structure fire, the placement and operation of initial attack lines protect occupants and fire fighters as well as providing the first water on the fire for extinguishment.
- The IC should conduct a risk versus benefit analysis before the decision is made to allow fire fighters to enter a building to conduct an aggressive interior attack.
- The effectiveness of an initial attack depends on several decisions that must be made in rapid succession.
- This decision making begins when the first company arrives on the fire ground and the company officer assumes command.
- A decision needs to be made as to the mode of operation used, offensive or defensive.
- A decision also has to be made regarding the size of hose line to be deployed during the initial attack, how many hose lines are needed, and where they should be positioned.
- The type of nozzle must also be considered, as it will determine the flow and shape of the stream.
- Spray nozzles operate with a stream pattern from straight stream to wide-angle fog and smooth-bore nozzles operate with a solid-stream pattern.
- During an offensive fire attack, an engine company must also consider whether to use a direct or indirect attack.
- A direct attack is one in which a solid or straight hose stream is used to deliver water directly onto the base of the fire.
- An indirect attack is one in which a solid, straight, or narrow fog stream is used to direct water at the ceiling to cool superheated gases in the upper levels of the room with the objective of preventing flashover, converting the water to steam, absorbing heat, and extinguishing the fire.
- Another important decision concerns the task of ventilating the building.
- Ventilation must be managed with suppression efforts, and a coordinated fire attack, supervised by the IC, will ensure that both tasks take place simultaneously.
- If the building is free burning, the building should be ventilated when the initial attack begins or as soon as possible thereafter.
- If it is a smoldering fire, however, the building must be ventilated before initial attack. In particular, a smoldering fire should be ventilated before any attempt is made to enter the fire building to prevent a backdraft explosion.
- After ventilated, a smoldering fire will burn freely and may be attacked using fire department standard operating guidelines.

Key Terms

Combination attack: A type of attack employing both the direct and indirect attack methods.

Direct attack: Firefighting operations involving the application of extinguishing agents directly onto the burning fuel.

Indirect attack: Firefighting operations involving the application of extinguishing agents to reduce the buildup of heat released from a fire without applying the agent directly onto the burning fuel.

1. When selecting the initial attack line:
   a. Select one of the preconnected hose lines.
   b. Select the largest preconnected hose line.
   c. Select the smallest preconnected hose line.
   d. Select either a preconnected hose line or hose line using hose from the hose bed.

2. An attack where water is directed in such a way as to create steam to absorb a large quantity of heat is a(n)
   a. Direct attack
   b. Indirect attack
   c. Absorption method
   d. Heat-reduction method

3. It is recommended that the 1½-inch hose line be
   a. Used only for fires contained to a small room
   b. Used only for all fires in residential buildings
   c. Used when the required rate of flow is 100 gpm or less
   d. Eliminated for structure firefighting

4. A 1¾-inch hose line can discharge _____ gpm.
   a. 60 to 125
   b. 120 to 200
   c. 125 to 150
   d. A maximum of 350

5. When a single 1¾-inch hose line is ineffective in extinguishing a larger fire
   a. Interior operations should be abandoned in favor of a defensive attack.
   b. Additional 1¾-inch lines should be deployed until the total quantity of water is sufficient to control the fire.
   c. Additional larger hose lines should be used.
   d. All of the above are options when faced with a situation where a single 1¾-inch line is insufficient.

6. Smooth-bore nozzles attached to the same size hose line
   a. Have greater reach than spray nozzles
   b. Will flow more water than spray nozzles
   c. Have greater reach and will flow more water than spray nozzles
   d. Are approximately equal to spray nozzles in reach and flow capacity

7. _____ stream(s) is/are the safest and most effective when conducting interior firefighting operations in an occupied area.
   a. Solid or straight
   b. Sixty-degree angle fog
   c. Wide-angle fog
   d. Intermittent straight and fog

8. A _____ stream is most effective in pushing the fire.
   a. Solid or straight
   b. Sixty-degree angle fog
   c. Wide-angle fog
   d. Intermittent straight and fog

9. When confronted with a well-involved fire in a part of a structure below an occupied area (e.g., basement storage area fire below apartments), it is usually best to
   a. Immediately rescue all visible occupants.
   b. Attack the fire using a 1¾-inch line with automatic nozzle, and lay a 1¾-inch line to the floor above the fire.
   c. Attack the fire using a 1¾-inch line with solid stream nozzle, and lay a 1¾-inch line to the floor above the fire.
   d. Attack the fire using a 2½-inch line with solid stream nozzle, and lay a 1¾-inch line to the floor above the fire.

10. Of the methods listed, advancing hose via _____ will require the most hose to advance a line to an upper floor of a structure.
    a. The stairs
    b. A ladder
    c. A rope
    d. A pike pole

11. A _____ fire must be ventilated before an offensive attack is begun.
    a. Free-burning
    b. Postflashover
    c. Smoldering
    d. All fires must be ventilated before an interior attack is made.

12. Backdraft is defined as
    a. The rapid ignition of all contents and fire gases in an area
    b. Fire gases suddenly igniting overhead
    c. A sudden explosive ignition of fire gases
    d. An oxygen-starved fire

Backup Lines

Chapter 7

Learning Objectives

- Realize that backup lines must be stretched whenever it is obvious that the fire will not be quickly extinguished with initial attack lines.

- Understand that backup lines should have greater reach and delivery than the initial attack lines.

- Recognize that if smaller initial attack lines are ineffective, they should be shut down and larger backup lines should be placed in operation.

- Examine the need to have adequate personnel on scene to operate initial attack lines and to also stretch and operate sufficient backup lines.

If one basic point were to be made in a discussion of initial fire attack, it would be to select the proper size hose and a nozzle to develop a stream of water that the situation demands, whether it be from a 1¾-inch hose line, 2½-inch hose line, or a master stream appliance. With the proper size streams, fire fighters stand a better chance of controlling and extinguishing the fire in a shorter period of time while preventing further property damage and protecting themselves.

Nevertheless, there is always the possibility of a sudden increase in fire activity after the initial attack has begun. A particularly flammable fuel, interior construction that does little to limit the spread of fire, flashover, or the spread of fire to a new fuel supply could quickly aggravate the fire situation to the point where the initial attack lines are incapable of extinguishing the fire. There are other such events that could occur on the fire ground—enough so that the engine company must always be ready to deal with them.

Backup lines are the engine company's first line of defense. A **backup line** is an additional hose line used to reinforce and protect personnel in the event the initial attack proves inadequate. They are hose lines to be used when initial attack lines cannot quickly control the fire. Backup lines are not used for exposure coverage or to attack the fire at other positions. Rather, they are held in readiness to back up the attack lines, in the same general area, when and if they are needed. This chapter addresses the positioning, sizing, and use of backup lines. It also describes personnel considerations. Backup lines should be deployed on all structural interests.

Positioning of Backup Lines

Backup lines should be stretched whenever it is obvious that the fire will not be quickly extinguished with initial attack lines. This means that backup lines should be taken into the fire building immediately after the initial attack lines. They should be positioned close to and set up to cover the same area as the initial attack lines. They should then be charged and placed into operation if needed.

Backup lines are of little use if they remain on the pumper. In the time it takes to get a backup line off the pumper, stretched into position, charged, and placed into operation, a fire could easily get completely out of control. Backup lines must be where they are needed at the time they are needed if they are to establish control of the fire.

Key Points

Backup lines are used when initial attack lines cannot quickly control the fire.

Key Points

Backup lines must be where they are needed at the time they are needed if they are to establish control of the fire.

Key Points

Backup lines should have greater reach and deliver more water than initial attack lines.

Sizes of Backup Lines

If initial attack lines are not gaining control of a fire, it is probable that the streams are not reaching into the seat of the fire or that these hose lines are not delivering enough water to cool the area. The use of additional hose lines of the same size will not solve the problem. They will certainly not penetrate any further into the fire than the original attack lines, and they will do little toward increasing the cooling effect.

For these reasons, the backup lines should have greater reach and deliver more water than the initial attack lines. This means that, in general, backup lines should be at least one size larger than initial attack lines.

Backing Up 1¾-Inch Lines

Although many fire departments still use 1½-inch hose lines for attack lines, for this discussion, 1¾-inch hose will be featured. When two 1¾-inch hoselines are used for initial attack, the backup line should be a 2½-inch hose line **Figure 7-1a**.

The 2½-inch hose line will deliver almost as much water with far greater reach than the two 1¾-inch handlines. It will allow deeper penetration into the fire and a lot more cooling—especially if a solid stream with a 1⅛ inch or 1¼-inch smooth-bore tip is used. A 2½-inch attack line should always be backed up by another 2½-inch hose line **Figure 7-1b**.

Adding another 1¾-inch attack line would do little except add to the water load in the fire building. If two 1¾-inch lines were being used and they were ineffective, three or four of them would be just as ineffective. The result would be more water damage, not more fire control. The penetration and cooling effect of the larger hose lines are what is needed.

Key Points

When two 1¾-inch hose lines are used for initial attack, the backup line is a 2½-inch hose line.

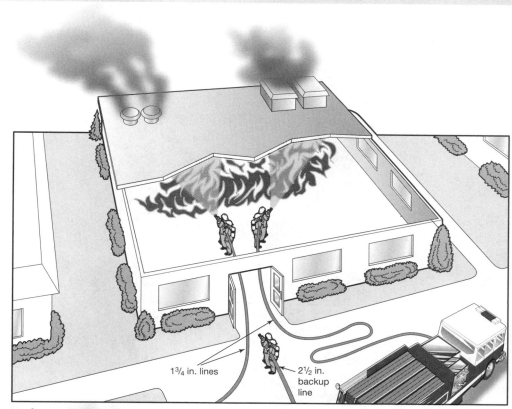

(a) 1¾ in. lines

1¾ in. lines

2½ in. backup line

(a) 1¾ in. attack lines

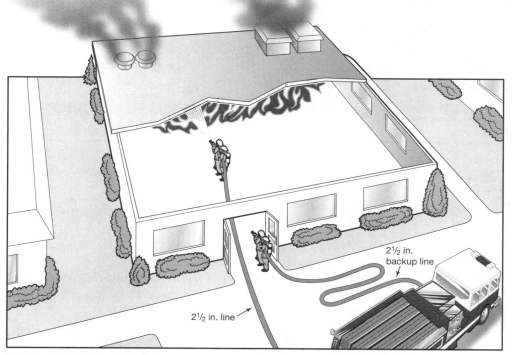

2½ in. backup line

2½ in. line

(b) 2½ in. attack lines

Figure 7-1 (a) The 1¾-inch attack lines are reaching but not penetrating the fire and the water volume is not sufficient to knock down or extinguish the fire. (b) The 2½-inch attack line is reaching and penetrating the fire with sufficient water volume to begin control. An additional 2½-inch hose line is used to back up the initial attack line.

Figure 7-2 One 2½-inch hose line may be sufficient to back up two or more 1¾-inch attack lines when they are operating within one or two floors of each other.

The fire situation and the location and number of 1¾-inch lines being used will determine how many 2½-inch backup lines are required. For example, one 2½-inch line may be sufficient to back up two or more 1¾-inch lines on the same floor if the floor is divided into apartments, work areas, or offices. When a large open area is involved, the number of backup lines might have to be increased. A 2½-inch backup line may be advanced to a particular floor to back up 1¾-inch lines on that floor and on the floor above **Figure 7-2**.

Backing Up 2½-Inch Hose Lines

If 2½-inch hose lines are used for initial attack, backup lines should be 2½-inch hose lines with larger tips unless these are already in use. Backing up such hose lines is discussed in the next section.

If the 2½-inch initial attack lines have spray nozzles, the backup lines should be equipped with smooth-bore tips.

Key Points

The fire situation and the location and number of 1¾-inch lines being used will determine how many 2½-inch backup lines are required. When a large open area is involved, the number of backup lines might have to be increased.

Key Points

If 2½-inch hose lines are used for initial attack, backup lines should be 2½-inch hose lines with larger tips unless these are already in use.

Key Points

A backup line with a smooth-bore tip will deliver more water, and the stream will penetrate further into the fire.

Most 2½-inch spray nozzles deliver up to 250 gpm. They can be adjusted from a wide-angle spray of almost 90 degrees to a straight stream; however, the water delivery rate cannot be increased over this maximum in any position.

A backup line with a smooth-bore tip will deliver more water, and the stream will penetrate further into the fire. This stream thus will have a better chance of controlling the fire than a straight stream from a spray nozzle and in the long run can gain effective control with less water expended.

In many departments, the 2½-inch hose line is the largest weapon available for attacking a fire. If a fire has gained considerable headway by the time the first pumpers arrive, these hose lines should be placed in service without delay.

Master Streams for Back Up

When a fire is so serious that the heaviest hose lines (2½-inch hose with 1⅛-inch and 1¼-inch tips) must be used for the initial attack, the chances of fire spread and increased intensity are great. It is imperative that master stream appliances be available for back up. Portable deluge sets, deck guns, ladder pipes, platform nozzles, and so forth should be set up ready for use if an adequate water supply is available. If the responding department does not have such equipment but a neighboring department does, then that department should be called to the scene as soon as the extent of possible involvement is determined.

The master stream appliance should immediately be charged, either to the device itself or to a control point close to the appliance to be placed in service with no loss of time. In many cases, master stream appliances have been necessary to cover the escape of fire fighters from particularly dangerous

Key Points

When a fire is so serious that the heaviest hose lines must be used for the initial attack, master stream appliances must be available for back up.

Key Points

A master stream appliance that is charged for back up can quickly be placed in service to cover the escape of fire fighters from dangerous positions.

positions, keeping away heat and fire as the crew made their way to safety. A master stream appliance that is charged for back up can quickly be placed in service for such use.

When an interior attack using 2½-inch hose lines fails to control the fire, back out all fire fighters, and attack using master stream appliances. This would call for changing tactics from an offensive mode to a defensive mode of operation.

When a fire requires the use of master stream appliances from the start, the problem is usually that of controlling overall fire spread. In this case, the need is not for backup lines, but rather for additional, heavier streams of the proper type for general fire attack. These operations are covered in detail in Chapter 9.

Use of Backup Lines

If all goes well, the initial attack lines will extinguish the fire, and the backup lines can be withdrawn; however, if the backup lines have to be placed in service, the initial attack lines should be shut down if they are ineffective against the fire. There is little that they can do in support of the heavier streams. An incident commander should not allow fire fighters to stand around operating ineffective attack lines while the building burns down around them. Ineffective handlines should be shut down, and a concerted effort made toward getting additional heavier streams into operation. After the backup lines have gained control of the fire and an advance can be made, the smaller lines can be

Key Points

If backup lines are placed in service, the initial attack lines should be shut down. The latter have already proved to be ineffective against the fire; there is little that they can do in support of the heavier streams.

Key Points

After the backup lines have gained control of the fire and an advance can be made, the smaller lines can be used to perform extinguishment and overhaul operations.

Figure 7-3 For final extinguishment and overhaul, 1¾-inch hose lines are efficient and effective.

used to perform fire extinguishment and overhaul (mop up) operations. You may redeploy fire fighters from the attack line to the backup line.

If a 2½-inch hose has been used for both initial attack and backup lines, then a 1¾-inch hose line can be attached to the 2½-inch nozzles or wyes for extinguishment and overhaul operations. Again, the purpose is to minimize the amount of water delivered into the building, consistent with complete and efficient extinguishment of the fire **Figure 7-3** .

Personnel

Hose lines are needed to attack the fire, to get above and around the fire, and to back up initial attack lines. These operations take adequate personnel. The backup assignment must not be overlooked, nor should it be given to fire fighters who have already been engaged in initial attack operations. These fire

fighters will be tired and beaten down by the physical activity, heat, personal protective equipment, and SCBA should be used during the initial fire attack. If they are pulled off the initial attack lines and assigned to backup lines, the backup operation will be understandably inefficient. Additional fire fighters must be assigned to backup duties. Moreover, backup lines should be stretched while initial attack is in progress. If this is done, it is impossible for the same crew to perform both duties. The

Key Points

The backup assignment should not be given to fire fighters who have already been engaged in initial attack operations. If there are not enough personnel at the fire scene, additional fire fighters should be summoned promptly.

incident commander is responsible for maintaining adequate personnel on scene to perform the necessary tasks that need to be carried out.

If there are not enough personnel at the fire scene, additional fire fighters should be summoned promptly. In volunteer departments, backup duties can be assigned to fire fighters who arrive after the first companies. Where this is impractical, neighboring departments can be called to handle this assignment. In paid departments with additional on-duty companies, there should be no hesitation in getting the necessary engine companies on the scene. Smaller paid departments should also plan for backup operations, even if arrangements must be made ahead of time with neighboring departments through a mutual aid agreement.

By the same token, master stream appliances needed for back up should be set up by incoming engine companies laying the proper supply lines to operate the appliances. This eliminates much hand and legwork on the part of fire fighters. Hand laying large-diameter hose or multiple supply lines is sometimes necessary. This action can be slow, consuming a lot of energy. The master stream appliances should be charged and manned by fresh crews, again for speed and efficiency. If fire fighters assigned to the initial attack are forced to withdraw from the fire building, they can hook up their 2½-inch hose lines into the master stream appliances for increased water supply if this has not already been done.

Key Points

Master stream appliances needed for back up should be set up by incoming engine companies laying the proper supply lines to operate the appliances. The master stream appliances should be charged and manned by fresh crews.

In addition to backup lines, a rapid intervention crew of at least four fire fighters equipped with full PPE, SCBA, and necessary tools and equipment should be standing by ready to enter the building. This crew consists of personnel who may be needed to respond immediately to assist in locating or removing a lost or trapped fire fighter or to any incident involving a fire fighter in distress.

Key Points

In addition to backup lines, an rapid intervention crew of at least four fire fighters equipped with full PPE, SCBA, and necessary tools and equipment should be standing by ready to enter the building.

Wrap-Up

Chief Concepts

- Backup lines should be larger than the initial attack lines in gallons per minute flow and should be laid out, charged, and assigned a crew to operate them if necessary.
- Use a smooth-bore nozzle on backup lines. Advance backup lines into the building in the same general area as initial attack lines.
- The crew assigned to the backup line must be in full PPE and SCBA so that they can quickly enter the building to assist other fire fighters if necessary.

Key Term

Backup line: An additional hose line used to reinforce and protect personnel in the event the initial attack proves inadequate.

1. Backup lines can be used to
   a. Augment an unsuccessful initial fire attack
   b. Exposure protection
   c. Attack the fire on the floor above the fire
   d. All of the above

2. Backup lines are needed
   a. At every fire where lines are placed inside the structure for fire control purposes
   b. Whenever it is obvious that the fire will not be quickly brought under control
   c. When the initial attack line proves to be insufficient
   d. When in a defensive strategic mode

3. Backup lines should
   a. Be at least as large as the initial attack line
   b. Be at least as large as the initial attack line and any other line being used to extinguish the fire
   c. Have a flow greater than the initial attack lines
   d. The size and flow capacity of a backup line is determined by actual and potential fire conditions.

4. When master streams are used to attack the fire from the exterior
   a. Change tactics from an offensive to a defensive mode
   b. Move fire fighters away from interior positions where the master stream is being discharged
   c. Coordinated interior and exterior operations for the greatest flow potential
   d. Close doors leading from common areas to the compartment where the master stream is being discharged

5. Backup lines should be staffed by
   a. The same company that is operating the initial attack line whenever possible
   b. A company other than the company that is operating the initial attack line
   c. Could be by either the first-arriving unit or later-arriving unit depending on the total staffing per company and the size of the attack and backup lines
   d. The rapid intervention crew

6. Use a _____ nozzle on backup lines.
   a. Spray nozzle
   b. Smooth bore
   c. Automatic
   d. The nozzle type depends on the situation.

Exposure Protection

Chapter 8

Learning Objectives

- Define the term exterior exposure fire as one that spreads from one structure to another or from an independent part of a building to another.

- Define the term exterior exposure and the factors that affect the severity of a fire during an incident involving an exterior exposure.

- Examine tactics used to control basement fires by preventing the vertical and horizontal spread of fire.

Firefighting operations can be difficult and dangerous for fire-ground personnel attempting to protect exposures. Exposure protection is needed to shield a building or a part of the building, which has been subjected to radiant and convected heat as well as direct flame impingement from the main body of fire. An adequate water supply must be initiated, and fire fighters must place hose lines in strategic locations to cover the maximum amount of exposed area. These positions should afford them protection from the effects of the fire while providing a vantage point to protect exposures and stop the spread of fire into uninvolved areas of a building.

As in any fire-ground activity, the incident commander (IC) must consider the size of the fire and the risks to fire fighters performing exposure protection operations. The IC must continually evaluate the current conditions to ensure the safety of fire fighters working in designated positions. This can be accomplished by using the risk versus benefits system. Fire fighters must work in teams, follow directions, and refrain from freelancing on the fire ground. They must pay attention to their surroundings and be cognizant of the fact that they are working in a dangerous environment.

Exposure coverage is second only to rescue on any list of the basic objectives of a firefighting operation. Structures near a fire building, the exterior exposures, and parts of the fire building not yet involved, the interior exposures, must be protected to minimize the danger to their occupants as well as to contain the fire. At many fires, the efforts of engine companies to develop an adequate water supply quickly, get initial attack lines in service, and back up those attack lines with other hose lines are necessitated by exposure problems.

As in every phase of firefighting, area company inspections and preincident planning are important parts of exposure protection. **Preincident planning** will help to locate exposure hazards (conditions or situations in or around a structure that will promote the spread of fire), as well as areas or neighborhoods in which fire spread is especially likely. Preincident planning should ensure that sufficient equipment and personnel are dispatched on the first alarm to cover these exposures. Where necessary, the first alarm response should be increased over normal assignments.

The adequate number of fire fighters is the key to full exposure protection. Implicit in all of the discussions in this chapter is the need for fire fighters and equipment over and above those engaged in fighting the fire itself. Fire fighters are needed to check for secondary fires, to control them, to direct hose lines

Key Points

Preincident planning will help locate exposure hazards, those conditions or situations in or around a structure that will promote the spread of fire.

Key Points

The adequate number of fire fighters is the key to full exposure protection.

on exposed structures, to set up additional water supply lines, to open up interior channels in which fire may be spreading, as well as carrying out other tasks that may need attention. Personnel and the necessary equipment must be available.

Responding ladder companies will perform some of these jobs; if they are not available, engine company personnel must do this work. Additional alarms may be needed to summon the desired resources to combat the fire. Mutual aid agreements between neighboring communities, agencies, and/or counties should also be established to ensure additional resources.

Any special equipment that will aid in firefighting should be dispatched to the fire scene on the first alarm. If a department does not have such equipment but a neighboring department does, arrangements should be made to have the neighboring department respond to the first alarm with the necessary equipment. If fire departments do not call for this equipment until they reach the fire scene and find out it is needed, the loss of time may have serious consequences. For example, if a fire department does not have an aerial ladder or additional mobile water supply apparatus available and they will be needed, they should be requested as soon as possible.

In this chapter, both exterior and interior exposures are discussed. The latter are the less spectacular but may demand as much effort from fire fighters to control. Basement fires, which can lead to serious interior exposure problems, are discussed in the final section of the chapter.

Exterior Exposures

The term exposure fire applies to the outside exposure; such a fire is regarded as one that spreads from one structure to another or from one independent part of a building to another, as across a court or between the wings of a building **Figure 8-1**. An "exposure hazard" is a condition that will promote the spread of fire if a fire should start in or reach that area.

Unpierced fire walls and spaces that check fire from spreading between buildings or stacked materials are the greatest deterrents to exposure fires and are of great assistance to fire

Key Points

An exposure fire is one that spreads from one structure to another or from one independent part of a building to another.

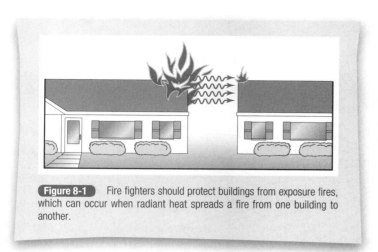

Figure 8-1 Fire fighters should protect buildings from exposure fires, which can occur when radiant heat spreads a fire from one building to another.

fighters when there are severe outside fires. Outside sprinklers and spray systems are also a great help, but unfortunately, they are rare items in fire protection equipment except in special installations.

Initial Response Considerations

Factors affecting the severity of an exterior exposure problem include the following:

- Recent weather
- Present weather, especially wind conditions
- Spacing between the fire and the exposures
- Building construction design and materials
- Intensity and size of the fire
- Location of the fire
- Availability and combustibility of fuel
- Size of the firefighting force
- Firefighting equipment on hand
- Availability of water

The worst combination of factors might be recent dry weather, strong winds blowing toward the exposures, an area of closely spaced frame buildings, a severe fire that is difficult to reach, plenty of easily ignited materials located between the fire building and exposures, and limited personnel and apparatus response to the first alarm.

Of these factors, the fire department normally has control of only the fire force and equipment responding to the first and additional alarms. For this reason, preincident planning is crucial.

Key Points

Many of the factors that affect exterior exposure problems may not be known before fire fighters reach the fire ground; therefore, equipment must be available when it is needed. Fire fighters must be trained in exposure coverage, and sufficient personnel must be available to perform all fire-ground tasks.

The fact that the department has control over only 2 of the 10 factors affecting exposure problems gives those two factors special importance. Equipment must be available when it is needed. Fire fighters must be trained in exposure coverage, and sufficient personnel must be available to perform all fire-ground tasks.

If preincident planning has been done properly and fire fighters have been trained, responding companies will be aware of the construction and spacing of buildings and the availability of fuel in and around the fire area. Sufficient personnel and equipment should be available and ready to respond. What happens in the first few minutes on the fire ground will dictate the end results over the course of the incident.

Recent weather will be a matter of record, and present weather must be observed.

What may not be known at the time of the alarm are the size and intensity of the fire and its location, either in the building or in the area. These factors will directly affect the amount of heat radiated from the fire building. The size and intensity of the fire, combined with the extent of structural involvement, will determine the amount of radiated heat. Radiant heat not only will keep fire fighters away from the fire building but will also add to exposure problems. **Figure 8-2** shows some of the characteristics of radiant heat.

Other factors that cannot be evaluated until the company arrives at the fire ground are wind direction and velocity. Hot air, smoke, gases, and embers will rise in convection currents and then will be carried downwind. The result could be a chain of exposure fires.

Convection Exposures

Flying firebrands and embers carried by convection and wind currents can cause serious fire containment problems. They are especially dangerous in lumber yards and other open storage areas. A severe building fire, especially one that is burning through the roof, can create strong convection currents above the building. In such situations, many fire departments have experienced exposure fires at great distances from the original fire, sometimes creating a problem worse than the initial fire **Figure 8-3**. There are areas of the country where wood shingle roofs exist. Wood shingle roofs can also contribute to the firebrand problem.

One way to guard against this possibility is by patrolling areas that are downwind from large fires. Under a Unified Command System, the police department and other authorized radio-equipped personnel, such as public works employees

Key Points

Radiant heat and convection may adversely impact exposure problems.

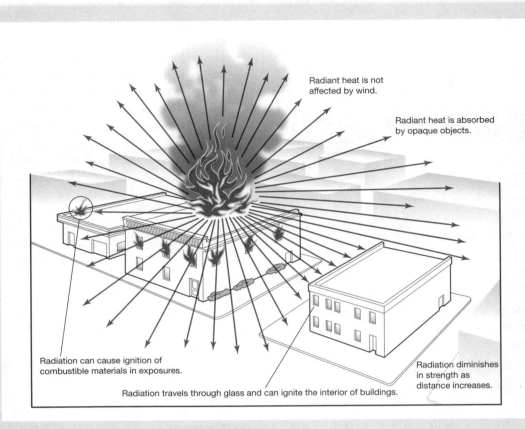

Radiant heat is not affected by wind.

Radiant heat is absorbed by opaque objects.

Radiation can cause ignition of combustible materials in exposures.

Radiation travels through glass and can ignite the interior of buildings.

Radiation diminishes in strength as distance increases.

Figure 8-2 Radiant heat travels in straight lines in all directions.

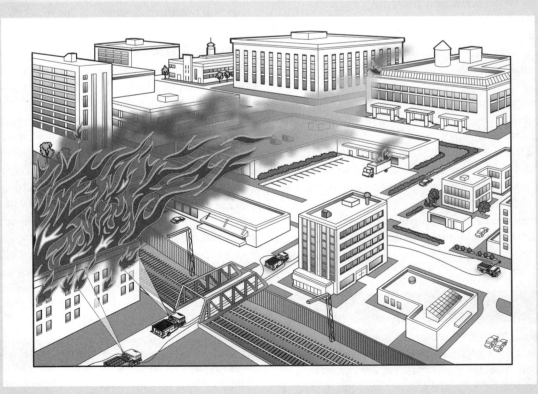

Figure 8-3 Convection can spread firebrands and embers to exposures some distance away.

Key Points

Exterior exposures must be protected from radiant heat and flying firebrands and embers that are carried by convection and wind currents.

or employees of local utilities, can be assigned to the patrols. Communications between the fire department and these personnel must be efficient if the patrols are to be effective. Training sessions between interagencies should be held to assure smooth operation when such patrols are required.

Patrols are extremely important when wind conditions are severe. A secondary fire that is located quickly can be extinguished before it develops into another source of wind-carried firebrands. If it is permitted to burn unchecked, such a fire could become as serious as the original fire in a very short time. Many companies may already be committed at the original fire, using large volumes of water. If a second serious fire develops close by, there may be insufficient apparatus, personnel, or water to control both fires. The potential for disaster is great.

Key Points

Patrols are extremely important when wind conditions are severe.

Key Points

The only way to protect exposures from radiant heat is to cool them by the application of water.

Radiant Heat Exposures

Radiant heat moves away from the fire building in all directions; winds do not affect it. Thus, fire may spread by radiation to any building near enough to the fire building to absorb sufficient heat. The only way to protect exposures from radiant heat is to cool them by the application of water.

Because water itself is transparent and radiant heat will pass through it **Figure 8-4**, operating a stream of water between the fire and the exposure will not protect the exposure. The radiant heat will move through the stream and heat the surface of the exposure to its ignition point. Instead, the stream must be directed onto the surface of the exposure in such a way that the water washes down its walls. The water will absorb heat from the exposure and thus keep it from igniting.

Radiant heat also will pass through transparent glass and ignite materials within a building. If the outside surface of a building is in danger of ignition from radiant heat, the areas within its windows constitute an equal hazard. In this case, the building should be entered, and each floor should be checked.

Hose lines should be taken into the exposure building so that they can be placed into operation if needed. If the building

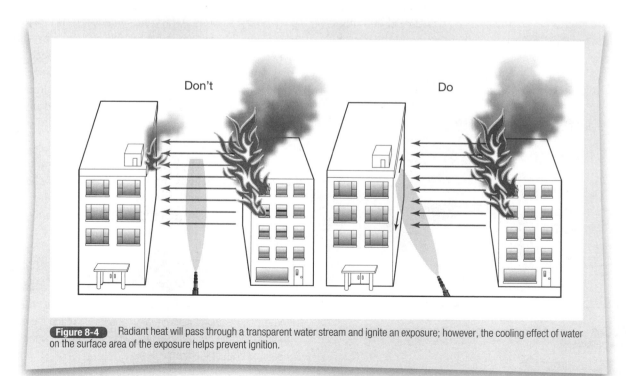

Figure 8-4 Radiant heat will pass through a transparent water stream and ignite an exposure; however, the cooling effect of water on the surface area of the exposure helps prevent ignition.

Key Points

The structures or the materials nearest the fire must be covered first, beginning on the leeward side of the fire.

Key Points

When exposures are fairly close to the fire building, the most vulnerable areas are the parts of the exposed buildings just above the fire where radiant heat, hot air, gases, and smoke tend to concentrate.

has a standpipe system, it should be charged. If the exposure has a sprinkler system, a pumper should be readied to charge the system if necessary. (The use of standpipe and sprinkler systems is discussed in more detail in Chapter 10.)

Exposure Coverage

The size and location of the fire and the wind conditions, factors not known until the company arrives at the fire ground, will determine what actions are to be taken in covering exposures. The structures or the materials nearest the fire must be covered, beginning first on the leeward or downwind. It is the combination of convected and radiant heat that makes the leeward side most dangerous. After the leeward side is covered, other areas must be protected because of the spread of fire by radiant heat alone.

When the exposures are fairly close to the fire building, the most vulnerable areas are the parts of the exposed buildings just above the fire **Figure 8-5**. The radiant heat and the hot air,

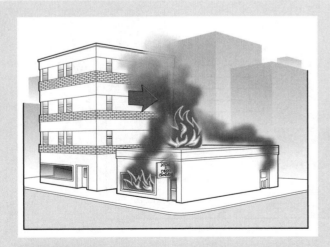

Figure 8-5 The point of greatest danger to this exposure is just above the top of the flames.

gases, and smoke will tend to concentrate there. To be effective, streams should be directed onto the exposure at a level somewhat above that of the fire.

Hose Lines and Nozzles

Water Streams As in fire attack, the hose lines and nozzles must be of the proper size and volume. To control the spread of fire and accomplish extinguishment, there also must be enough of them. Streams developed by nozzles should deliver enough water to cool the exposure and should be large enough to reach the exposure, penetrating the fire if necessary, without being dissipated by the heat and draft of the fire. The streams must carry through the fire area to the exposure with enough water to prevent ignition. In some instances, one fire fighter with a 1¾-inch line can protect an exposure; in others, 2½-inch hose lines or master stream appliances may be required.

Although hand-held hose lines can be effective, they require more staffing, place fire fighters closer to the building, and provide less water than master streams. Circumstances may occur where a hand-held hose line is the only way to protect an exposure, but master streams provide a safer and more effective stream for most situations requiring exposure protection.

Streams must also be large enough so that they are not overly affected by winds. A strong wind may necessitate the use of streams that are larger than usual, especially on the windward side of the fire.

Streams developed by spray nozzles, although effective in cooling exposures, are very susceptible to break up and reduction of efficiency by winds. Streams developed by smooth-bore nozzles hold together better and therefore will penetrate winds better than spray-nozzle streams. If the winds are too strong for spray-nozzle streams, the heaviest smooth-bore streams might have to be used for exposure protection, including those produced by master stream appliances. Streams produced by smooth-bore nozzles can be powerful and could cause damage to buildings and the appendages. Streams can break glass on the exposure, making it more vulnerable for fire extension within the building.

Hose Line Positions The positioning of exposure lines is especially important. Exposure lines must be placed where they will cover the maximum amount of exposed area. If one stream will not cover the exposure completely, then additional streams should be used **Figure 8-6**. It is advantageous to position

Key Points

Streams developed by nozzles should deliver enough water to cool the exposure and should be large enough to reach the exposure, penetrating the fire if necessary, without being dissipated by the heat and draft of the fire.

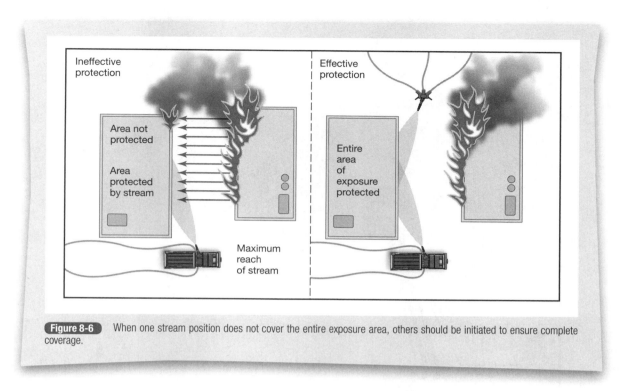

Figure 8-6 When one stream position does not cover the entire exposure area, others should be initiated to ensure complete coverage.

a stream so that the exposure is protected while the stream is brought to bear on the part of the fire nearest the exposure; however, concern for hitting the fire should not be allowed to reduce the effectiveness of exposure coverage.

The ideal position is one that maximizes the reach and effectiveness of the stream while providing fire fighters protection from radiant heat and keeping them out of the collapse zone. Adjoining roofs, secure and safe stacks of noncombustible stored materials, and buildings across alleys, courts, or narrow streets often make good positions for exposure lines. Companies should be aware of such locations; onsite training sessions are a great help in developing a plan of attack prior to an actual incident.

Figure 8-7 shows various vantage points in protecting exposures. Hose lines should be brought into exposed buildings, and protective systems in both fire and exposed buildings should be supplied.

Interior Exposures

The objective of interior exposure coverage is the same as that of exterior exposure coverage: to keep the fire from spreading to uninvolved areas. Interior exposure fires could require the use of

hose lines in many positions inside the building to stop the spread of fire. The mobility of smaller hose lines, such as the 1¾ inch, makes them desirable for interior exposure coverage, but on occasion, the severity of interior fire spread may necessitate the use of larger hose lines, such as the 2½ inch.

Whereas outside exposure fires are obvious and easily seen, many interior exposure fires are not at all obvious; they must be sought out and located by fire fighters on the scene. If fire is spreading up stairways, through halls and corridors, or up elevator shafts, there is a very good chance that it also is spreading vertically and horizontally through walls, ceilings, and other concealed spaces. Performing these tasks properly and safely requires adequate personnel. Many structure fires have resulted in the complete and total destruction of the building because fire fighters were unable to get ahead of the fire in concealed spaces because of inadequate staffing levels. The fire is not going to take a time out while the IC musters adequate staffing to combat the fire.

Fire in Concealed Spaces

Although there are signs that indicate to fire fighters that fire is spreading within a concealed space, there are no signs to indicate

Key Points

The best exposure line position is one that maximizes the reach and effectiveness of the stream while providing fire fighters protection from radiant heat and keeping them out of the collapse zone.

Key Points

Interior exposure fires could require hose lines in many positions inside the building to stop the fire from spreading to uninvolved areas.

Figure 8-7 There are several vantage points in protecting exposures.

that fire has not spread to a concealed space. If there is any possibility of fire in a horizontal or vertical space or shaft, that space must be opened and inspected visually. If necessary, streams must be directed into the concealed space, and it must be ventilated.

These actions will cause damage to the building, but there is little choice in the matter: Either open up shafts, walls, partitions, ceilings, floors, or whatever or let the fire destroy the building completely. Although every effort should be made to minimize damage to the building and its contents, openings have to be large enough for inspection, hose line manipulation, and ventilation activities. The openings also must be able to take in enough water to extinguish the fire.

The opening of concealed spaces and ventilation outlets is ladder company work and should be assigned as such; however, if there are no ladder companies at the fire scene or if they are not a part of the available fire force, engine company personnel will have to do the job. Ladder company operations are performed at every fire, regardless of who does them.

Vertical Fire Spread

Fire will travel vertically inside walls and partitions and through pipe shafts, dumbwaiters, air shafts, and similar pathways in a

---

Key Points

Many interior exposure fires are not obvious; they are concealed. If fire is spreading up stairways, through halls and corridors, or up elevator shafts, there is a very good chance that it also is spreading vertically and horizontally through walls, ceilings, and other concealed spaces.

Key Points

If there is any possibility of fire in a horizontal or vertical space or shaft, that space must be opened and inspected visually.

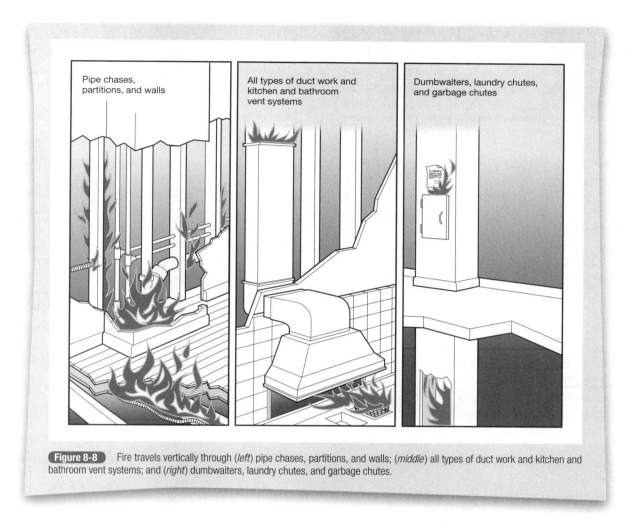

Figure 8-8 Fire travels vertically through (*left*) pipe chases, partitions, and walls; (*middle*) all types of duct work and kitchen and bathroom vent systems; and (*right*) dumbwaiters, laundry chutes, and garbage chutes.

building **Figure 8-8**. Many structures contain concealed vertical shafts that carry building utilities such as water, gas, and electric lines or sewer system vent pipes. Many single-family dwellings and apartment houses have central heating system vents that extend through the building from the basement to a chimney fixture on the roof.

These vertical channels are normally placed toward the rear of commercial buildings, stores, and shopping centers. In apartment buildings, they follow the pattern of the apartment layout. They are most often found near the kitchens and bathrooms, and each shaft is usually placed so that it serves two, four, or more apartments. The great variety in the design of single-family dwellings means that vertical shafts could be located almost anywhere in these structures; however, the locations of vent pipes and kitchen vents on the roof are good indications of where these shafts can be found **Figure 8-9**.

Signs of Vertical Spread If there is a working fire inside a building, fire fighters should assume that flames have entered concealed spaces until they determine otherwise. As they arrive at the fire ground, engine company personnel should be looking

for signs that fire has extended into vertical channels within the building.

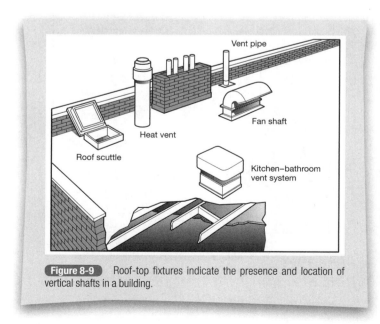

Figure 8-9 Roof-top fixtures indicate the presence and location of vertical shafts in a building.

Key Points

Indications of vertical fire spread seen outside buildings include smoke or fire showing around roof features and the condition of the tar on roofs and around roof features.

Key Points

Indications of vertical fire spread inside buildings include smoke or flame issuing from walls, blistering or discoloration of paint or other wall coverings, and a wall that is hot to the touch.

Smoke or fire showing around roof features such as vent pipes is an obvious sign that fire has spread to the shafts leading to these outlets. Close examination may show that the tar on some parts of the roof is soft and shiny; this means that the tar has been melted slightly by heat in a shaft below it. In rain or snow, a clear or dry spot around a roof feature over a shaft may indicate that heat from the fire is in the shaft or that fire itself has entered the shaft.

Inside the building, such signs as smoke and flames issuing from walls and blistering or discoloration of paint or other wall coverings will indicate the presence of fire within concealed shafts, walls, or partitions. A wall that is hot to the touch probably is concealing fire. If any of these signs is found, the wall must be opened Figure 8-10.

The thermal imaging camera has become an invaluable tool for use on the fire ground. By capturing heat images, the device is able to show the relative temperature of different objects on a display screen. One advantage of the camera is that it can locate fire and heat in concealed spaces. The device is able to locate the source of fire behind walls, ceilings, and other voids. It can also indicate the direction of spread.

Control of Vertical Spread　If it is known or suspected that the fire has entered a vertical shaft, hose lines should be directed into the shaft, and it should be opened and inspected from the roof. Because fire in these channels, although intense, is confined to a comparatively small area, 1¾-inch hose lines will be most useful for this operation. In addition, the mobility of these hose lines permits them to be moved rapidly from one location to another with minimum personnel.

Fire will travel vertically as long as it is allowed to do so. Opening the roof will encourage vertical travel and slow the horizontal movement of the fire Figure 8-11. In a sense, the fire will be "led" out of the structure and away from cocklofts and attics, spaces between floors and ceilings, and other horizontal

Figure 8-11　Vertical ventilation is essential when a fire is burning in an attic or cockloft to keep the fire contained and limit horizontal movement of fire.

Discoloration
Blisters on wall surface
Hot to the touch
Smoke pattern at moldings

Figure 8-10　When there are signs of fire in walls, partitions, and vertical shafts, the wall should be opened. The fire can then be attacked with the proper size attack line.

Key Points

Fire will travel vertically as long as it is allowed to do so. Opening the roof encourages vertical travel and slows the horizontal movement of the fire.

channels. Opening a skylight in an attic may not ventilate the cockloft area if it is enclosed. If a vertical shaft does not reach to the roof, the roof opening should be made directly above the shaft or as close to that point as possible. A hose line should be advanced to the attic to attack any fire that has spread there.

To attack the fire from below, an opening should be made in the wall or vertical space that shows signs of fire. An opening that is about waist high will allow a fire fighter on one knee to handle a nozzle easily and to direct the stream up into the opening. An opening at this level also allows the nozzle to be directed down if the fire is coming from below.

It is important that at least one stream be directed upward toward the fire to control embers, hot gases, and smoke, as well as the fire itself. If using a spray nozzle, adjust the stream pattern to provide maximum reach and coverage within the shaft or partition. If it is evident, because of the height of the building, that this one hose line will not suffice, more hose lines must be stretched above or below to ensure full coverage of the exposure area. If the fire is traveling in several separate but parallel shafts or between the studs in walls and partitions, the area should be opened completely, again to ensure full coverage.

In a large vertical opening, such as an elevator shaft or a stairway, it may be necessary to use 2½-inch attack lines to control the fire because of the volume of fire that a large vertical opening will support. This will be especially true when several elevators are arranged in a bank in one shaft or where stairways are unusually wide.

Horizontal Fire Spread

Although fire tends to travel vertically, it will also travel horizontally through any available paths. Fire may travel horizontally through the spaces between ceilings and floors, over false or hanging ceilings, along ductwork and utility conduits, through conveyor tunnels in industrial buildings and warehouses, and through

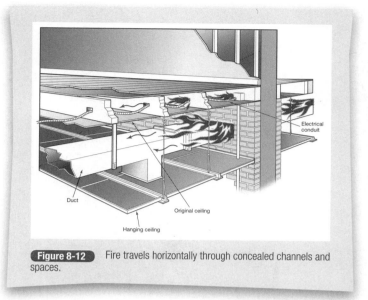

Figure 8-12   Fire travels horizontally through concealed channels and spaces.

similar channels in other buildings. In addition, construction features may cause concealed horizontal channels to be formed within walls, floors, and ceilings. Such channels can permit fire to spread horizontally through the building.

Fire may also move horizontally from one building or occupancy to an adjoining one through ducts, ceiling spaces, and walls **Figure 8-12**. An example would be the spread of fire in a row of stores or from one apartment to another.

Exposure lines therefore must be positioned inside the fire building as well as in adjoining units. Ceilings, floors, and other horizontal paths suspected of or showing signs of fire travel must be opened for inspection and, if necessary, fire attack operations initiated. For the most part, there are few exterior signs of fire spread through horizontal channels, except when the flames reach and involve exterior walls. The interior signs are the same as the signs of vertical fire spread listed in the previous section.

In general, horizontal exposures are controlled in the same way as vertical exposures. The channel must be opened up, and an adequate stream must be directed at the fire. Property conservation should be executed during this task to cover or remove valuable property, thus keeping damage and loss to a minimum. The most important goal, however, is to control the fire **Figure 8-13**.

Key Points

Fire may travel horizontally through the spaces between ceilings and floors, over false or hanging ceilings, along ductwork and utility conduits, through conveyor tunnels in industrial buildings and warehouses, and through similar channels in other buildings. Often these channels are concealed.

Key Points

In general, horizontal exposures are controlled in the same way as vertical exposures. The channel must be opened, and an adequate stream must be directed at the fire.

Figure 8-13   A check for horizontal fire spread should be made on each side of the main body of fire.

Open Interior Spread

One other interior exposure problem can often be serious, although the exposure is not concealed. The situation occurs when an interior fire has gained control of a large enclosed area, such as a supermarket, warehouse, or other large enclosed space that includes little or nothing to deter the spread of fire. The problem is essentially that of a large outside exposure; however, the fire is constrained by the walls and roof of a building. Although the fire may not be endangering adjoining structures, the intense heat and the smoke and gases contained by the building will pose severe problems for fire fighters.

The IC must conduct a risk versus benefit analysis before committing fire fighters to interior structural firefighting operations. In this situation, fire attack operations might be more closely related to exterior firefighting than interior, even though the fire is inside. Large handlines and even master stream

Key Points

An open interior fire in a large enclosed area, such as a warehouse, poses a severe problem for fire fighters. This situation may create intense heat, smoke, and gases because nothing within the building deters the fire, which is constrained by the walls and roof.

appliances might be needed inside the structure to cover exposed areas and to knock down the fire. Moreover, in such a building, the roof can become a problem in terms of both firefighting and safety. If interior operations cannot be conducted in a safe manner, fire fighters should be withdrawn from those positions and the operation changed from an offensive to a defensive mode.

The unprotected steel roof of a typical supermarket is a good example. The rising heat and hot combustion products of a major fire could cause the roof supports to buckle and the roof to collapse. Frequently, a large fire may have to be controlled before interior exposures can be protected. Until the fire is extinguished, the possibility of roof collapse makes the interior of the store extremely hazardous to fire fighters. Collapse may occur in little as 10 minutes with bar joist roofs.

If it is safe to do so, the fire should be rapidly and thoroughly ventilated through the roof, as close over the main fire area as possible. This could be a great help in slowing the tendency of the fire to spread through the enclosed area.

If the involved occupancy adjoins other buildings, they should be checked for fire spreading through ceiling spaces and other channels. If the fire is large and numerous exposures exist, it will be necessary to handle two kinds of exposure problems at the same time. One will require big hose lines and larger flows of water; the other will require smaller, fast-moving hose lines.

Basement Fires

Basement fires are one of the most difficult and dangerous types of incident that fire fighters will encounter. There may not be many entry points into the basement area from either the inside or outside of the building. Ventilation openings may also be few in numbers. First-arriving company members must size up the situation quickly and determine the location and probable size of the fire, points of entry, and construction features of the building.

The IC should conduct a risk versus benefit analysis to determine whether the area is safe to enter. Offensive fires should be conducted from the interior, unburned side of the building. Unfortunately, the exact location of the fire and its travel direction may be difficult to ascertain because of smoke and high heat conditions. When fire is showing out of basement windows or doorways, an aggressive interior attack should be made from the unburned side. In most of these incidents, the fire, heat, and smoke are usually venting out in the proper direction (the burned side).

If fire is burning up stairways or other vertical openings into the first floor, an attack line must be positioned to protect the integrity of the stairway. Ventilation is a key element in controlling the fire by permitting the products of combustion to escape while allowing fire fighters to gain entry into the basement. Sufficient attack lines and backup lines must be provided by an adequate number of fire fighters to accomplish the objective. Fire fighters should avoid operating streams from opposite directions. A coordinated fire attack must be ensured.

One of the objectives is to keep the fire from extending to upper floors. The IC should be aware of extension of the fire vertically into upper floors, including the attic and cockloft areas. Fire fighters need to be assigned above the fire to open up any vertical or horizontal space that is suspected of containing fire. Hose lines must be positioned and operated by engine companies if these conditions are encountered. Fire fighters must keep ahead of the fire at all times. With this said, consider the following information on basement fires.

The higher a fire is in a building, the better off fire fighters and occupants are in terms of fire spread. A fire in the basement of an apartment house is much more dangerous to occupants throughout the building than a fire on an upper floor. Heat, smoke, gases, and hot embers traveling vertically through the building can ignite secondary fires and overcome occupants on all floors.

Control of Basement Fires

When the fire is at the very bottom of the structure and forcing its way up into the building, it is important to control the main body of fire as quickly as possible. All openings into the basement should be located and considered as points of attack. Doors, windows, access chutes, trapdoors, and any other openings should be used for positioning hose lines to attack the fire. Cellar pipes and/or nozzles should be placed in service in positions that will permit them to knock down or assist in extinguishing the fire. Standard attack lines of the proper size should be used where possible. Obviously, outside attack lines are not to be used from openings into the basement, such as windows and doors, if fire fighters are working in the basement.

Along with the application of sufficient streams, a basement fire requires full ventilation of both the basement and the first floor. Basement windows, preferably opposite those being used for fire attack, should be used for ventilation. The first floor should be ventilated completely to allow efficient and safe operations. Moreover, forcing the products of combustion out of the building through the basement and first floor reduces the chance that they will create problems on upper floors **Figure 8-14** .

Key Points

When the fire is at the very bottom of the structure and forcing its way up into the building, it is important to control the main body of fire as quickly as possible.

Figure 8-14 Available openings to the basement can be used for attack from one direction while ventilation is being conducted from another.

Key Points

Along with the application of sufficient streams, a basement fire requires full ventilation of both the basement and the first floor.

Basements serving mercantile buildings can be heavily loaded with product, often stacked up to the ceiling. This limits the effectiveness of streams. In such situations, additional hose lines will be required to control the fire, and thorough ventilation of the basement and first floor will be extremely important.

In residential or commercial buildings that have basement doors to the outside, it is best to use these doors for fire attack and to ventilate through basement windows; however, interior stairs to the first floor must be covered. Fire must be attacked in these areas, and the first floor must be vented. It might be efficient to attack the fire down interior stairs.

Where outside entrances are available but there are no windows to the basement, the situation becomes more difficult. Solid streams on larger attack lines may be needed. The fire may travel vertically more quickly, and the first floor will be in greater peril. Observing the effect of initial attack streams especially will

be important. If the attack appears to have a quick effect on the fire, ventilation of the first floor may suffice. If not, openings should be made in the first floor just inside windows; then the windows should be removed or opened to provide ventilation of the basement, improve fire attack effectiveness, and lessen interior exposure problems.

In fighting a basement fire in a store, the storefront and display window area should be used for fire attack or ventilation according to the fire situation and the presence of other basement openings. If there is a rear door but no other openings to the basement, the attack can begin from that position, and ventilation efforts should be conducted through openings at or near areas already burned **Figure 8-15**. Where there are other openings such as windows or sidewalk doors, opening of the storefront might not be necessary. If the basement has no outside opening but only an interior stairway or a trapdoor entrance, the display

Key Points

In residential or commercial buildings that have basement doors to the outside, it is best to use these doors for fire attack and to ventilate through basement windows.

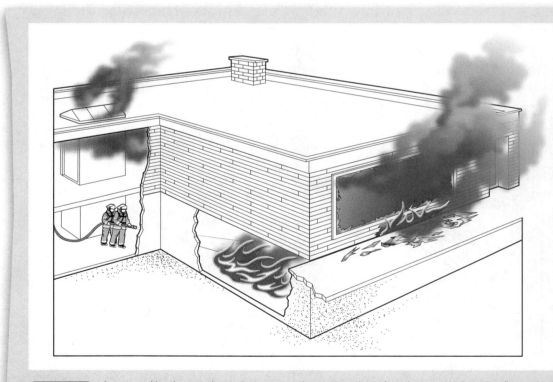

Figure 8-15   In a store with only a rear door to the basement, the attack can begin from the rear, and the wall or floor under the front display window can be opened for ventilation.

window openings can be used for ventilation, and the fire can be attacked with hose lines through interior openings or with cellar nozzles.

In a basement fire in a store, the storefront and display window area should be used for fire attack or ventilation according to the fire situation and the presence of other basement openings.

Protecting Exposures

Stairways and other openings from the basement to the first floor are major channels for the vertical spread of a basement fire. These must be covered as soon as possible. Attack lines should be brought in to first-floor landings to knock down any fire that may have spread from the basement; walls and partitions should be checked for vertical fire spread, and baseboards should be removed in suspect areas.

These first-floor operations are extremely important, but the safety of fire fighters and occupants is even more essential. When a basement fire has gained considerable headway or fully involves the basement, the IC must make a decision whether fire fighters should enter the building. If a decision has been made to allow entry into the building, fire fighters must pay particular attention to conditions within the structure. The floor should be checked for a soft or spongy feeling. Smoke coming through the floor should be noted, especially at outside entrances, along walls, and at other places that may involve main structural members. These

signs indicate a weakening of the floor. If conditions deteriorate, fire fighters should be removed from the building immediately.

Vertical Spread   Many of the vertical openings in a building originate at the basement level, and thus, there is nothing to keep fire from entering the openings and spreading through them. Vertical openings should be opened at the first floor if there are any signs of fire spread there. In any case, these openings should be opened at roof level to allow the heat, smoke, and gases to leave the building. Otherwise, it may be possible to have a fire in the basement and one at the top of the building, with nothing in between.

Horizontal Spread   Adjoining basements must be checked for horizontal fire spread. Hose lines must be taken to those locations to extinguish extending fire. This is particularly important in older structures, where deterioration, remodeling, and additions might have destroyed the integrity of the walls between buildings. Party walls, those that support beams or joists from two buildings, often have openings where the beams rest on the walls. Fire can easily spread from basement to basement through these openings. Regardless of their construction, adjoining basements must be checked.

The same is true for adjoining attics, especially if the fire building was not vented quickly. Whether the fire originated in the basement or in some other area of the building, the tops of adjoining buildings must be checked and, if necessary, vented.

Wrap-Up

Chief Concepts

- Exposure protection is initiated to protect a building or part of the building, which has been subjected to radiant and convected heat as well as direct flame impingement from the main body of fire.
- On the fire ground, it is imperative that ample personnel and equipment are available to cover exposures while the main body of fire is being attacked.
- An adequate water supply must be obtained and sufficient hose lines placed in strategic locations to cover the maximum amount of exposed areas.
- Large fires present a difficult situation because of the amount of radiant and convected heat created by the fire.
- Directing hose lines on exposures will keep combustible material from reaching its ignition temperature.
- During large fires, the use of master stream devices can deliver a significant amount of water onto the exposure from a safe distance.
- An exterior exposure is a structure, or other object, to which the main body of fire can extend.
  - It may be a separate structure or an independent part of the main fire building, such as a separate wing or a building separated by a courtyard.
- Radiant and convected heat, embers, and wind currents must be considered when protecting exterior exposures.
- Interior exposures are areas within the main fire building.
  - These exposures are protected from radiant and convected heat and direct flame impingement.

- Vertical and horizontal channels or voids such as walls, ceiling, and subfloors must be opened to expose and extinguish any fire that may be traveling in these areas.
- A basement fire presents a difficult and dangerous situation to fire fighters because the fire might be hard to reach and the entire building may be exposed to the spread of fire.
- A basement fire must be managed using a coordinated fire attack.
- Ventilation must be conducted early, and without it, fire attack may be difficult if not impossible due to high heat and low visibility.
- Engine companies must use the safest entry point to attack the fire and sufficient hose lines must be available to accomplish the task of extinguishing the fire.
- Fire fighters must pay particular attention to their surroundings, maintain orientation within the area, and have a secure means of egress.

Key Term

Preincident plan: A written document resulting from the gathering of general and detailed information to be used by public emergency response agencies and private industry for determining the response to reasonable anticipated emergency incidents at a specific facility.

1. Exposure coverage is _____ on any list of the basic objectives of a firefighting operation.
   a. The primary concern
   b. Second only to rescue
   c. Third behind rescue and extinguishment
   d. Fourth behind rescue, confinement, and extinguishment

2. _____ is the key to full exposure protection.
   a. Properly placed apparatus
   b. An adequate number or apparatus with pump capacity
   c. An adequate water supply
   d. An adequate number of fire fighters

3. When wind and convection currents carry fire brands away from the area of origin to distant locations
   a. Additional fire units should be staged at locations downwind and patrol assigned sectors to extinguish quickly any fires caused by fire brands.
   b. Police should be used to patrol the areas where fire brands are likely to spread fire.
   c. Police and other authorized radio-equipped personnel should be used to patrol the areas where fire brands are likely to spread fire.
   d. A secondary line of defense, well beyond the fire zone, should be established downwind.

4. _____ heat is affected by wind.
   a. Convected
   b. Conducted
   c. Radiant
   d. All three of these heat transfer methods are affected by wind.

5. When protecting exposures from a radiant heat source, it is best to
   a. Place a fire stream between the fire building and exposure
   b. Push the fire away from the exposure using a wide-angle fog pattern
   c. Direct water onto the surface of the exposed building
   d. All of the above are effective means of protecting exposures depending on the fire intensity, wind conditions, and other factors.

6. Glass remaining in a window frame
   a. Will not protect against radiant heat
   b. Will not protect against convected heat
   c. Will not protect against flying brands
   d. Will provide some measure of protection from all three heat sources listed previously

7. In a commercial building, concealed vertical shafts carrying building utilities are most likely to be found
   a. On the front wall near utility entry points from the street
   b. In the center of the building above basement heating equipment
   c. Toward the rear of the building
   d. Almost anywhere

8. Vertical ventilation is essential
   a. At all fires
   b. At all residential fires
   c. At all fires in buildings with attics or cocklofts
   d. When a fire is burning in the attic or cockloft

9. In terms of fire spread, fire fighters and occupants are better off when the fire
   a. Is located on the lowest level of the building
   b. On the top floor
   c. On an intermediate floor
   d. Each of the fire locations listed above present different problems for occupants and fire fighters. Although different, the risk is essentially the same.

10. A basement fire
    a. Is normally impossible to ventilate
    b. Should be vented on the wall opposite the fire (or as far away from the fire as possible) when vent openings are available on at least two sides of the basement
    c. Should be vented on the wall closest to the fire
    d. Should be vented in the basement and first floor whenever possible

Master Stream Appliances

Chapter 9

Learning Objectives

- Identify the three types of master stream appliances and their application on the fire ground, including the advantages and disadvantages of each.

- Recognize the various types and sizes of nozzles used to deliver the proper flow rates and stream patterns when using a master stream appliance.

- Recognize that the effectiveness of a master stream appliance depends almost completely on its receiving an adequate supply of water.

- Understand how a master stream appliance can be used to the best advantage to control and extinguish a fire.

<u>Master stream appliances</u> are large-caliber devices that are used primarily during defensive operations. Conditions on the fire ground may be deteriorating or have already reached that point when these appliances are placed in service. During this time, fire fighters must be completely aware of their surroundings on the fire ground.

Master stream appliances mean that a large volume of water is being applied on the fire ground. Fire fighters must use extreme caution working around hose lines, pumpers, and the appliance itself. Precautions must be observed around buildings or other structures in which the appliance is being directed.

Both pumpers and aerial fire apparatus should be positioned so that radiant heat, convective heat, burning embers, or structural collapse will not affect the safety of fire fighters working in those areas. Master stream appliances should be as far away from the fire as possible while still being effective. If there is any indication of structural integrity, a perimeter should be set up a safe distance from the collapse zone and all personnel and equipment kept out of this area.

Command must manage the use of master stream appliances on the fire ground. Master stream appliances should not be placed in service in a defensive mode while an offensive, interior attack is being conducted. Serious injury to fire fighters could result as fire, heat, and products of combustion are forced onto them. In addition, elevated master stream appliances should not be operated into vertical natural openings or openings created by fire fighters or the fire itself. This blunder will certainly endanger fire fighters inside the building and prevent the fire, heat, and gases from leaving the building.

Master stream appliances, also known as "heavy-stream" and "large-caliber stream" devices, deliver more water and can reach further than the largest hand-held hose lines. They are placed into operation when hand-held hose lines are ineffective in fire attack, for exposure protection, and for backup lines. Several types of appliances are available, with several sizes of nozzles, including both spray nozzles and smooth-bore tips.

Master stream appliances are not special equipment. They should be considered standard firefighting tools and, as such, should be part of preincident planning and training evolutions. To be effective, these appliances require water flows from 350 to 2,000 gpm, depending on the size and type of the nozzle being used and the volume of water available. Engine company personnel should be fully trained in the operation of these appliances, from the laying and charging of supply lines to the

Key Points

Master stream appliances are used when hand-held hose lines are ineffective; they are used in fire attack, for exposure protection, and for backup lines.

Key Points

There are three types of master stream appliances: portable appliances, fixed appliances, and elevated master stream appliances.

use of the appliances in fire attack and exposure protection. This chapter discusses the types of appliances and nozzles that are available, water supply for master streams, and the use of master stream appliances at the fire ground.

Types of Master Stream Appliances

There are essentially three types of master stream appliances. The first two types are portable and fixed appliances, which are carried on and operated from the fire apparatus. In addition, portable appliances can be operated on the ground or from other remote positions. The third type, elevated master stream appliances, is operated from aerial fire apparatus ladders or platforms. Each has its advantages and disadvantages in a particular fire situation.

Portable Master Stream Appliance

Portable master stream appliances are often referred to as deck guns or monitors. They are carried on the apparatus and generally are operated from there, although they are designed to permit operation from either the apparatus or the ground. A number of varieties are available, but all are operated in similar fashion.

Portable master stream appliances can be operated from a fixed position on the pumper and placed into service quickly. If the pumper can be placed in a strategic location so that the appliance can be operated onto the fire or to protect an exposure, then it may be advantageous to work from that position. If the pumper cannot be located near the fire, the appliance could be removed from the pumper and operated remotely from the truck.

Portable master stream appliances are provided with points on the bottom of the stabilizing legs, as well as a chain or straps. The points dig into the ground while the chain and straps are attached to an unmovable object to prevent the appliance from shifting during operation.

When water flows through the appliance, the nozzle reaction force created can cause the device to move, or shift, from its position. This can occur if the nozzle is operating at too low an angle. A safety lock is provided so that the lock must be manually released before the nozzle can be lowered below a 35-degree angle. There are portable appliances that employ a safety shut-off valve that automatically shuts off the flow of water if the appliance moves. This feature reduces the risk of injury to fire fighters from an out-of-control appliance. Some portable master stream appliances are made to take advantage of a hose loop to prevent movement. When so equipped, this feature should be used.

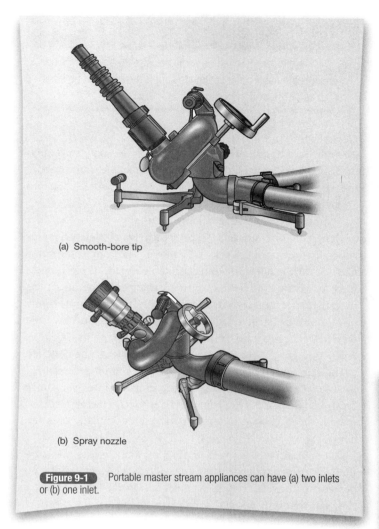

(a) Smooth-bore tip

(b) Spray nozzle

Figure 9-1 Portable master stream appliances can have (a) two inlets or (b) one inlet.

Figure 9-1 shows two types of portable master stream appliances with one and two inlets. These inlets are designed to be supplied by a large-diameter hose (LDH) or smaller diameter supply lines. They are capable of operating with either smooth-bore tips or spray nozzles.

Today's modern fire apparatus usually is equipped with a prepiped deck gun that can be quickly supplied from a separate discharge gate on the pump or taken off the apparatus, placed into a mounting bracket, and operated on the ground or other remote location. They may be equipped with a telescoping feature, which lowers the device for storage and raises it when operating to give greater clearance from other equipment including raised cab roofs. In addition, an appliance also may be remotely controlled from its position on the pumper. Engine company members must be familiar with the manufacturer's operating instructions for use from either the apparatus or the ground. A portable master stream appliance is placed on the ground and supplied with water from one or more hose lines.

Fixed Master Stream Appliances

Master stream appliances also can be permanently mounted or fixed to pumpers **Figure 9-2** . Water is supplied to fixed appliances in one of two ways. In the first method, water is prepiped to the appliance from a separate discharge gate on the pump. In the second method, the appliance is supplied directly by hose lines with one or more connections to the pumper's discharge outlets. Devices supplied by the second method also can be mounted on ladder trucks, special service vehicles designed to carry such appliances, and other vehicles without pumps. The apparatus on which a fixed master stream appliance is mounted must be carefully and correctly positioned at the fire scene.

Elevated Master Stream Appliances

Elevated master stream appliances **Figure 9-3** are found on aerial ladders, elevating platforms, and water towers. Unless the aerial fire apparatus is equipped with a fire pump, an engine company usually will supply water to the appliance.

Where a prepiped waterway is provided on an aerial ladder, the waterway system must be capable of flowing 1,000 gpm

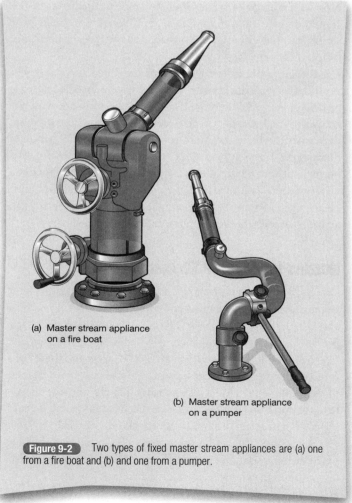

(a) Master stream appliance on a fire boat

(b) Master stream appliance on a pumper

Figure 9-2 Two types of fixed master stream appliances are (a) one from a fire boat and (b) and one from a pumper.

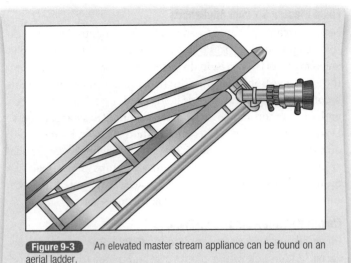

Figure 9-3 An elevated master stream appliance can be found on an aerial ladder.

at 100-psi nozzle pressure at full elevation and extension. A permanently attached monitor must be provided with a 1,000-gpm nozzle. Where a prepiped waterway is not provided, a ladder pipe with clamps to secure the appliance to the aerial ladder should be provided. In addition, appropriate tips, hose, hose straps, and halyards shall be provided to operate the appliance properly.

Elevating platforms of 110 ft or less rated vertical height must have a permanent water delivery system installed capable of delivering 1,000 gpm at 100-psi nozzle pressure with the elevating platform at its rated vertical height. One or more permanently installed monitors with nozzles capable of discharging 1,000 gpm must be provided on the platform. The permanent water system must supply the monitor. Permanent waterways on both an aerial ladder and elevating platform must be arranged so that it can be supplied at ground level through an external inlet that is a minimum of 4 inches.

Nozzles for Master Streams

Various sizes of smooth-bore tips are available for use with master stream appliances. The most common are the 1⅜-, 1½-, 1¾-, and 2-inch tips. These tips are all normally operated at a nozzle pressure of 80 psi. The 1⅜-inch tip will discharge 500 gpm, the 1½-inch tip 600 gpm, the 1¾-inch tip 800 gpm, and the 2-inch tip 1,000 gpm at 80-psi nozzle pressure. When water is delivered to any of these nozzles at a less-than-sufficient rate, the stream tends to break up. This decreases the reach of the stream. If water supply is a problem, a smaller tip size should be used.

A 1¼-inch tip can be used on a master stream appliance when there is not enough water volume to feed the tip sizes listed here. Although generally considered an attack line tip (325 gpm at 50 psi), it will develop a fairly heavy stream with a flow of

400 gpm at a nozzle pressure of 80 psi. A 1¼-inch tip is often furnished as part of a ladder pipe's assortment of smooth-bore tips. Tips of this size could also be used on other master stream appliances when necessary.

Spray nozzles are available in many sizes. They are designed to operate at 100 psi, with water delivery rates generally from about 300 to 1,250 gpm. There are some spray nozzles with greater flow rates available. Again, these nozzles must receive a sufficient supply of water and operate at the proper nozzle pressure in order to be effective. Some spray nozzles are constructed so that their flow rate can be varied or selectable, generally in increments of 250, 350, 500, 750, and 1,000 gpm. Allowing the volume of water to be matched to the capability of the supply system, they are recommended for use in areas where water supply delivery rates vary from location to location. Others are available with a factory ordered fixed orifice, whereas others may have an automatic pressure control.

It should be obvious from the previous discussion that a master stream appliance is only as good as its water supply; no device can be effective with an inadequate flow of water. A water flow rate that is inadequate for a particular size of smooth-bore tip, however, might be just right for a smaller smooth-bore tip. Moreover, a 500-gpm stream from a 500-gpm tip is much more effective in controlling a fire than an inadequate stream from a 1,000-gpm tip. For this reason, it is important that smooth-bore tips of several sizes be carried on the apparatus for use as required.

Water Supply for Master Stream Appliances

Master stream appliances operate at high flow rates, which increase friction loss in supply lines and thus require higher pressures and increased water flow from pumpers. As explained earlier, the effectiveness of a master stream appliance depends almost completely on it receiving an adequate supply of water; therefore, it is important that friction losses be minimized and that pumpers be used most efficiently. For this reason, the following points are recommended to supply a master stream appliance:

- Locate a pumper at a hydrant or other water source capable of flowing a volume of water sufficient to supply a master stream appliance on the fire ground.
- Use an LDH or an adequate number of smaller supply lines laid between pumpers if engaged in relay pumping operations.
- If the master stream appliance is to be operated away from the pumper, use an LDH or an adequate number of smaller supply lines to maintain adequate volume to the appliance.
- Maintain a minimum distance between the pumper and master stream appliance if smaller supply lines are being used. The use of an LDH will usually allow distances between the pumper and master stream appliance to be increased significantly.

Before discussing these recommendations individually, consider the following three scenarios. In the first two scenarios, 2½-inch hose lines are being used to supply the master stream appliance from a pumper, whereas the third scenario uses a 4-inch supply line. For these examples, the pumper and the master stream appliance are 100 ft apart **Figure 9-4**.

Scenario 1

In the first scenario, a pumper is supplying a 1¼-inch smooth-bore tip, the smallest used on master stream appliances. This tip requires a nozzle pressure of 80 psi, which will develop a 400-gpm stream. This flow can be carried by a single 2½-inch hose line. Friction loss is approximately 35 psi per 100 ft; therefore, the pump pressure required for the 100-ft lay is 115 psi:

Nozzle pressure	80 psi
Friction loss	35 psi
Pump pressure	= 115 psi

Scenario 2

The second scenario has a pumper supplying a 750-gpm spray nozzle on the master stream appliance. The spray nozzle requires 100 psi. Two 2½-inch supply lines, with 375 gpm flowing in each, are being used. Friction loss is approximately 30 psi per 100 ft; therefore, the pump pressure required for the 100-ft lay is 130 psi.

Nozzle pressure	100 psi
Friction loss	30 psi
Pump pressure	= 130 psi

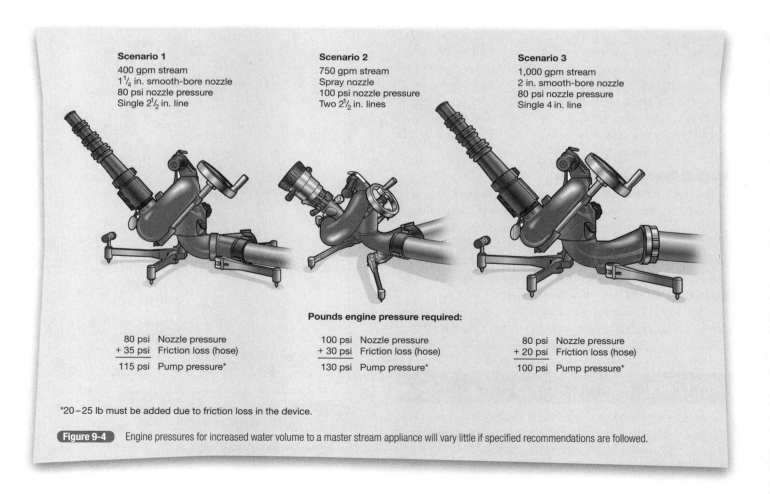

Scenario 1
400 gpm stream
1¼ in. smooth-bore nozzle
80 psi nozzle pressure
Single 2½ in. line

Scenario 2
750 gpm stream
Spray nozzle
100 psi nozzle pressure
Two 2½ in. lines

Scenario 3
1,000 gpm stream
2 in. smooth-bore nozzle
80 psi nozzle pressure
Single 4 in. line

Pounds engine pressure required:

80 psi	Nozzle pressure
+ 35 psi	Friction loss (hose)
115 psi	Pump pressure*

100 psi	Nozzle pressure
+ 30 psi	Friction loss (hose)
130 psi	Pump pressure*

80 psi	Nozzle pressure
+ 20 psi	Friction loss (hose)
100 psi	Pump pressure*

*20–25 lb must be added due to friction loss in the device.

Figure 9-4 Engine pressures for increased water volume to a master stream appliance will vary little if specified recommendations are followed.

Scenario 3

In the last example, a pumper is supplying a 2-inch smooth-bore tip on a master stream appliance. This tip requires a nozzle pressure of 80 psi, which will develop a 1,000-gpm stream that is being carried through a 4-inch supply line. Friction loss is approximately 20 psi per 100 ft; therefore, the pump pressure required for a 100-ft lay is 100 psi.

Nozzle pressure	80 psi
Friction loss	20 psi
Pump pressure	= 100 psi

Very little difference exists among the three calculated pump pressures, although different gpm flows were achieved using different size hose lines, nozzles, and nozzle pressures. If a single 4-inch supply line had been used for all scenarios, it would have been able to supply the master stream appliances over longer distances with negligible friction loss.

Pumper-to-Pumper Operation

As noted in Chapter 5, the most effective operation for delivering a large volume of water to a fire is to have pumpers at the hydrants or other adequate water source discharge their water to pumpers at the fire ground. This allows the use of a short lay of hose from the pumper at the fire to the master stream appliance with the friction loss minimized **Figure 9-5**. If an engine company operates a master stream appliance from the apparatus, it too should be supplied by a pumper at an adequate water source, whether or not the appliance is permanently connected to its pump. Supplying a pumper from another pumper may not be necessary if an LDH is used from an adequate water source. Many fire departments with adequate water delivery systems use 5-inch LDH to further reduce friction loss. A disadvantage of relay pumping is that two pumps are required to supply water to a master stream device.

Adequate Number of Supply Lines

In the three scenarios, the number of hose lines or their size between the pumper at the fire and the master stream appliance was increased when the required flow rate increased. This actually lowered the friction loss and kept the engine pressure fairly constant. If hose lines with less carrying capacity had been used in the last two examples, the friction loss would have increased to make the desired performance impossible.

The inlet of a master stream appliance is actually a large siamese, a device that collects water from two or more hose lines

Key Points

The most effective operation for delivering a large volume of water to a fire is to have the pumper at the hydrant or other adequate water source discharge its water to a pumper at the fire.

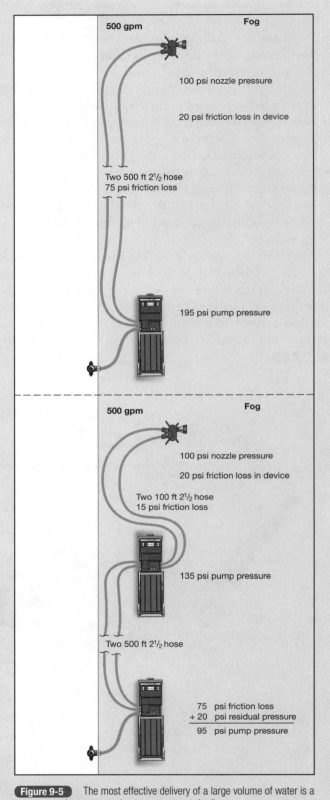

Figure 9-5 The most effective delivery of a large volume of water is a pumper at a hydrant pumping to a pumper at the fire.

Figure 9-6 A typical siamese connection combines two hose lines into a single line.

and delivers it to a single line, which in this case is the nozzle **Figure 9-6**. The inlet can also be a single LDH connection. More water meeting less resistance flows through the multiple lines or a single LDH line to reach the appliance. Thus, less pump pressure is required and a more efficient operation is carried out.

The 2½-inch hose was used in the first two examples only because it is the smallest supply line. A larger diameter hose would give even better performance because of greater carrying capacities and lower friction loss. There are master stream appliances manufactured with a single inlet, which is designed for use with LDH. This hose is discussed in Chapter 5.

It is suggested that a maximum length of 100 ft of hose be used between the pumper and a master stream appliance when using smaller supply lines. This recommended maximum is intended to limit the friction loss in the hose lines between the pumper at the fire and the appliance. Occasionally, there will be situations in which the master stream appliance must be more than 100 ft from the pumper at the fire because the pumper cannot be brought to the position necessary for proper operation. In such cases, it is important to remember that the longer hose lines will require greater pressures. Again, the use of LDH will allow greater carrying capacities with less friction loss than smaller supply lines.

Standard Procedures

Standard operating guidelines (SOGs) differ from jurisdiction to jurisdiction. A fire department must establish an SOG that reflects its specific needs predicated on the water supply available,

pumping capacities, and the water delivery system, including supply hose and appliances.

One other benefit can be derived from following the four recommendations given earlier: They are the basis of an SOG for placing master stream appliances into service. The recommendations themselves specify the size and length of hose to be used and the operation of pumping from one pumper to another. This combination is an effective one, as it minimizes pump pressures and thus maximizes water delivery capability.

The recommendations determine the discharge pressure at which the pump at the fire should be set, at least initially. The required pump pressure was close to the same in all three examples—about 115 to 130 psi. The friction loss in most master stream devices is 20 to 25 psi, which must be added to the engine pressures calculated in the examples. Each department must initiate a set of guidelines predicated on their current equipment and capabilities.

In most cases, a bit too much pressure will be no problem; there are no fire fighters trying to steady and hold a master stream appliance; however, a great deal of extra pressure will cause a solid stream to break up as soon as it leaves the nozzle, thereby making it ineffective. Its reach will be greatly reduced, and it will be as vulnerable to winds as a fog stream. In effect, too much pressure will turn a solid stream into a poor fog stream, a heavy spray of water accomplishing little if anything in extinguishing a fire or protecting an exposure. Fortunately, these results of too high of a pressure are obvious to a trained fire fighter observing the stream and a call for a reduction in engine pressure will solve the problem quickly.

Use of Master Stream Appliances

As noted earlier, several sizes of smooth-bore tips and spray nozzles should be carried on the pumper so that fire fighters can choose the correct nozzle for the fire situation and the water supply. It is just as important that both spray nozzles and smooth-bore tips are available for use as the situation demands.

A master stream appliance may be used for fire attack or exposure protection or to back up an existing hose stream. Most often, the appliance will be positioned on the outside to deliver water into the fire building through windows or doorways or to

Key Points

The use of an LDH will allow greater carrying capacities with less friction loss than smaller supply lines.

Key Points

Several sizes of smooth-bore tips and spray nozzles should be carried on the pumper so that fire fighters can choose the correct nozzle for the fire situation and the water supply. Both spray nozzles and smooth-bore tips should be available for use as the situation demands.

protect an exposure. Wind conditions and the distance to the fire or exposure will determine which type of nozzle is used and how it should be operated.

Solid-Stream Nozzles Versus Spray Nozzles

Because of the wind conditions mentioned, the strong draft created by a large fire and the usual distance from the nozzle to the building, the solid stream has proven to be most effective for fire attack, when using a master stream appliance. It will penetrate further into the building, covering more area of the fire.

For exposure protection, a spray nozzle using a fog stream may be superior to a solid stream if it is not affected by the wind or the distance from the fire building. The fog stream will cover a wider area than a solid stream and require less movement.

A spray nozzle using a straight stream, even from a master stream appliance, loses its effectiveness if it must be applied

over any distance **Figure 9-7** . If there is any question about the ability of a straight stream to reach a fire or exposure, a solid stream should be used. For example, the intensity of a fire and/or the structural integrity of the building might prevent the positioning of master stream appliances near enough for straight streams to be effective. In such a case, solid-stream nozzles with appropriate size tips would have to be used. Strong crosswinds will adversely affect both types of streams but can render a fog stream completely ineffective. Again, in a strong wind, a solid stream should be used. Smooth-bore nozzles are powerful and can damage an exposure building and its appendages. Glass can be broken, making the building more vulnerable for fire extension within the building.

Positioning the Master Stream Appliance

As stated earlier, to be effective, a spray nozzle using a straight stream must be positioned closer to the fire structure. Solid-

Figure 9-7 Straight stream from the master stream appliance (*left*) is breaking up from the effects of wind. This adverse effect may be avoided by moving the appliance closer to the fire building if conditions allow (*right*).

Key Points

Spray nozzles using a straight stream must be positioned closer to the fire structure. Solid-stream nozzles perform as well or better if they are positioned some distance from the building.

stream nozzles, on the other hand, will perform as well or better if they are positioned some distance from the building. For example, a solid stream may be used to attack a fire that is three or four floors above the level of the master stream appliance. For this operation, the appliance must be placed some distance from the building to achieve the proper angle of entry. If the appliance is too close to the building, the angle will be too steep. The stream then might not reach the fire area but rather flow water needlessly into an uninvolved area.

A properly positioned solid stream will enter the building at an angle that will cause the stream to be deflected over a wide area when it strikes the ceiling or other overhead, as shown in **Figure 9-8**. The low angle of stream 1, just over the sill into the building, is best for maximum penetration but may be affected by obstructions. The angle of stream 2 is also effective but may not penetrate as far into the building as stream 1. Stream 3 is ineffective because of limited penetration; therefore, a stream should be repositioned up and down in a window, doorway, or other opening to achieve the best combination of effective results.

Directing a Master Stream

To be most effective in fire control, a master stream should be moved horizontally back and forth across the fire area it is covering. The stream also should be moved up and down so that it reaches to the full depth of the fire area. The amount of movement depends on the extent of the fire and existing conditions. Although nothing may burn directly under a stationary stream, the fire could spread away from it. Movement of the stream assures coverage of a good-sized area in and around the fire **Figure 9-9**.

Less movement might be required for exposure coverage, especially when fog streams are being used. There may be times

Key Points

To be most effective in fire control, a master stream should be moved horizontally back and forth across the fire area it is covering. The stream also should be moved up and down so that it reaches to the full depth of the fire area.

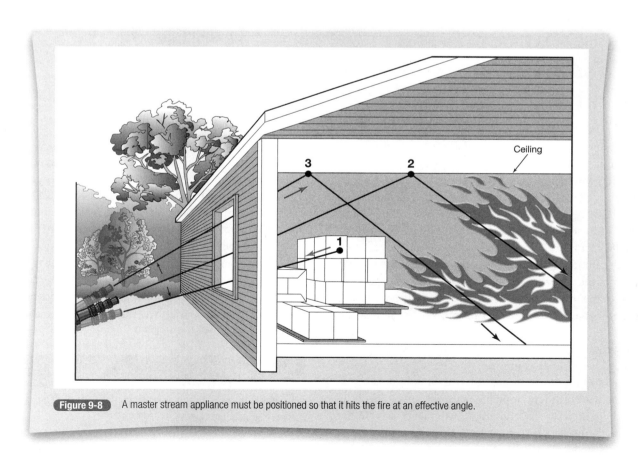

Figure 9-8 A master stream appliance must be positioned so that it hits the fire at an effective angle.

Figure 9-9 Master streams must be moved both vertically and horizontally to cover the entire area involved.

when a master stream appliance initially will be set up and left unattended. This may be due to a shortage of personnel or a situation where it may be unsafe for fire fighters, such as the possibility of a building collapse or a hazardous materials incident; however, a master stream appliance never should be set up and then left unattended indefinitely. The stream and its effect on the fire or the exposure should be carefully monitored. Command will need to be kept aware of the current conditions, as the appliance may need to be relocated to perform properly. If fire breaks out on an exposure, the exposure stream immediately may have to be moved to control the fire.

If a stream does not seem to be having any effect on a fire, it could be positioned improperly. The water must reach the base of the fire, and the master stream appliance must be positioned accordingly. Sometimes the effectiveness of the stream can be increased by moving the appliance so that the stream hits the fire at a better location. If repositioning is not

the problem, the size of the stream may need to be increased, or additional streams may need to be directed onto the fire.

In heavy smoke, it can be difficult to determine whether a stream is entering the building. Fire fighters directing the stream might not be able to see the windows. If possible to do so safely, an officer visually should check the building from a safe location. If this is not feasible, he or she should listen for the sound of the stream hitting the building and should look for heavy water runoff. Both of these signs indicate that the stream is not entering the building. The stream then should be adjusted until both

Key Points

If a stream does not seem to be having any effect on a fire, it could be positioned improperly.

Figure 9-10 Steam and white smoke indicate that the water stream is having an effect on the fire.

indications have disappeared **Figure 9-10**. At that point, the stream should be operating effectively.

Shutdown

A stream from an appliance should be used only as long as fire is visible in the area covered by the stream. A check should be made for visible fire. Steam and/or white smoke is an indication that the main body of fire in that area has been knocked down. When steam and/or white smoke no longer are visible, the fire apparently has been put out. The master stream then should be shut down or moved to cover another area; its work here has been completed. Continued operation at this point would only add to the water load on the fire building and the strain on the water supply system. If necessary, hand-held hose lines can be used for final extinguishment.

Elevated Master Streams

The same rules of operation discussed for portable and fixed master stream appliances apply for elevated master stream appliances. Because they are elevated above ground level, they can be used for fires on upper stories of buildings. They can be directed through windows or other openings by placing the nozzle at or near the window during defensive firefighting operations. When operating in a defensive mode, the chance of a building collapse must be considered. It is best to position the

aerial apparatus at the corners of the building or at a distance away from the building that will afford protection for fire fighters and the apparatus. This should safeguard them from radiant and convective heat, burning embers, and building collapse. They are also useful in the protection of exposures, especially on upper floors. Directing water on the exposure is more effective than directing a stream between the fire and the exposure.

A solid stream is most effective for fire attack when using a master stream appliance. The stream will penetrate further into the building covering more area of the fire. For exposure protection, a spray nozzle using a fog stream may be superior to a solid stream if it is not affected by the wind or the distance from the fire building. The fog stream will cover a wider area than a solid stream and require less movement. Several sizes of smooth-bore tips and spray nozzles should be carried so that fire fighters can choose the correct nozzle for the fire situation and the water supply. It is just as important that both spray nozzles and smooth-bore tips are available for use as the situation demands.

Elevated streams must never be directed into natural openings, such as skylights, scuttles and hatches, or holes made in the roof for venting or where fire has burned through the roof. Fire fighters vent roofs to allow smoke, heat, gases, and fire to escape the building so that an interior attack can be carried out. Vertical ventilation allows fire fighters to perform a primary search, get to the seat of the fire, and apply water to extinguish the fire. During an offensive mode of operation, it is a cardinal rule in the fire service that no attack line or master stream appliance is to be directed into a ventilation opening. This will only spread the smoke, heat, gases, and fire downward throughout the building, making it untenable for victims as well as fire fighters engaged in firefighting operations. If it is necessary to protect the roof from fire coming through these openings, a stream of water should be directed onto the roof adjacent to the opening so that the water will not flow into the opening. During a defensive operation when a major portion of the roof or large area is destroyed by fire, elevated master streams can be directed at visible fire inside the building without the danger of reversing ventilation.

Key Points

A master stream should be used only as long as fire is visible in the area covered by the stream.

Key Points

Elevated stream must never be directed into natural openings or holes made in the roof for venting or where fire has burned through the roof.

Wrap-Up

Chief Concepts

- Master stream appliances are placed in service when the streams from hand-held hose lines will not be effective in fire control or in exposure protection or as backup lines. They are primarily used during defensive operations.
- Engine company personnel should be well trained in the operation of these appliances. This includes ensuring that a proper water supply is available as well as selecting the proper supply line or lines to provide a sufficient water supply to operate the appliance safely and effectively.
- The most effective methods for supplying a master stream appliance involve the use of two pumpers; one at the water supply and one at the fire.
- Both spray nozzles and smooth-bore tips, in a range of sizes and styles, are available for use with these appliances. The size and style of the nozzle will depend on the fire situation and the available water supply.
- Effective operation of master stream appliances is mainly a matter of observing stream performance. The appliance should be positioned so the stream reaches into the fire and should be operated only as long as the stream is doing its job.
- When the fire is extinguished, the appliance should be shut down. If the stream has little effect on the fire, it should be replaced either with a larger stream or supplemented with additional streams. The basic objective is maximum use of the available water supply.

Key Term

Master stream appliance: A large-capacity nozzle that can be supplied by two or more hose lines or fixed piping and can flow in excess of 300 gallons per minute. It includes deck guns and portable ground monitors.

1. Master stream appliances should not be placed in service in a(n) _____ mode while an offensive, interior attack is being conducted.
   a. Offensive
   b. Defensive
   c. Marginal
   d. Interior

2. Master stream appliances _____ hand-held hose lines.
   a. Deliver more water from a greater distance than
   b. Deliver more water but must be placed closer to the target building than
   c. Deliver more water than most
   d. All of the above could be true depending on the size and type nozzle being used.

3. Master streams require water flows from _____ gpm.
   a. 200 to 1,200
   b. 500 to 1,500
   c. 500 to 2,000
   d. 350 to 2,000

4. A prepiped master stream appliance (deck gun) mounted on a pumper would be classified as a(n)
   a. Portable master stream appliance
   b. Fixed master stream appliance
   c. Elevated master stream appliance
   d. Either a fixed or portable master stream appliances depending on whether it can be removed from the apparatus

5. Elevated master stream appliances on _____ apparatus must have prepiped waterways.
   a. Aerial
   b. Elevating platform
   c. Both aerial and elevating platform
   d. Neither aerial or elevating platform

6. The minimum size smooth-bore tip for a master stream appliance is
   a. 1 inch
   b. 1¼ inch
   c. 1½ inch
   d. 1¾ inch

7. A portable master stream, with a 2-inch smooth-bore tip attached and a nozzle pressure of 80 psi is flowing at 1,000 gpm. The pumper supplying the master stream is 500 feet away and using a 4-inch hose as a supply line. What is the required discharge pressure for the engine supplying the portable master stream?
   a. 120 psi
   b. 140 psi
   c. 160 psi
   d. 180 psi

8. The _____ stream has proven to be most effective for fire attack when using master stream appliances.
   a. Solid
   b. Narrow-angle fog
   c. Wide-angle fog
   d. All of the above are effective. The stream to be used is determined by conditions at the scene including distance, wind, and angle.

9. Strong crosswinds will adversely affect the _____ stream.
   a. Solid
   b. Narrow-angle fog
   c. Wide-angle fog
   d. All of the above

10. Master streams operating from the exterior to the interior of a building should
    a. Remain in a stationary position providing the greatest penetration angle
    b. Be moved from side to side horizontally
    c. Be moved up and down vertically
    d. Be moved both horizontally and vertically

11. When operating a master stream from the exterior into a smoke-obscured window, the operator can determine whether the stream is entering the window by
    a. Observing windows that can be seen and estimating the distance between windows to determine the correct stream direction
    b. Listening for water striking a wall
    c. Observing the water runoff
    d. Answer choices b and c are both correct.

12. Elevated master streams should _____
    a. Never be directed into ventilation openings made in the roof
    b. Only be directed into roof ventilation openings after the attack becomes defensive
    c. Be directed into roof openings only when visible fire is showing at the opening
    d. Be cautiously directed immediately above visible fire coming from a roof opening

Fire Protection Systems

Chapter 10

Learning Objectives

- Examine the types and classes of standpipe systems and their purpose of providing water throughout a building for firefighting operations.

- Understand the equipment and tactics used to sustain an attack effectively and safely on a fire in a structure containing a standpipe system.

- Explore the use of a standpipe system in situations of exposure protection and for fire attack in adjoining buildings.

- Examine the types of sprinkler systems and the impact a system has in effectively protecting a building and its contents from fire.

- Consider other protective systems that use various agents to extinguish fires in buildings or in other process and/or uses.

Fighting fires in large buildings such as high-rise apartments or office buildings and large, complex buildings, such as warehouses or manufacturing facilities, can present a challenge to fire fighters. The safety of fire fighters is paramount when confronted with an incident of this scope and magnitude. To assist fire fighters in combating fires in buildings, a built-in fire protection system may have been installed. A standpipe system and/or a sprinkler system when properly designed, installed, and maintained can assist the fire department members if they are properly trained in its use. Firefighting operations should be based on a thorough knowledge of the building and the installed system. This knowledge is based on company inspections and preincident planning. In addition, a fire department must have written standard operating guidelines in place and an incident management system that will provide for the safety of fire fighters working in these hostile environments.

For fire fighters to work safely under fire conditions in buildings provided with a built-in fire protection system, they must know the type of system installed and how it functions. In addition, members must ensure an adequate water supply to the system so that fire control and extinguishment can be accomplished. If a system is not functioning properly, they must be able to address the problem and provide an alternative.

Command must have a planned fire attack, and the accountability of members, as in any fire situation, is a requirement that must be maintained throughout the incident. A sufficient number of fire fighters must be present and possess the proper tools and equipment to accomplish assigned tasks.

If elevators are installed in the fire building, fire fighters should following standard operating guidelines regarding their use. The incident commander must always conduct a risk versus benefit analysis before allowing fire fighters to use an elevator in a fire situation. The physical condition of fire fighters must be maintained during the incident, and fire fighters engaged in firefighting activities will need to be rehabbed and monitored in a timely manner.

Standpipe systems and sprinkler systems are both built-in fire protection systems, but they are very different in operation. Standpipe systems are usually installed in large structures such as high-rise apartment and office buildings and warehouses to eliminate the need for long stretches of hose lines from the street to the fire floor. They are designed to provide fire fighters with water supply outlets close enough to any fire to allow quick fire attack and extinguishment. Standpipe systems are installed in buildings for use by fire fighters and in some cases by building occupants.

An automatic sprinkler system, on the other hand, is designed to sense and apply water to a fire in its earliest stages. As the name implies, it does so initially without the help from fire fighters or building occupants. Fire fighters ensure that the water supply is maintained and the system is functioning properly while in operation. Both types of systems are described in this chapter, with the use of standpipe systems discussed in some detail.

Standpipe Systems

A **standpipe system** is a piping arrangement that carries water vertically and sometimes horizontally through a building for firefighting operations. The system's purpose is to provide a means of getting water to a fire without long, time-consuming hose stretches. For the most part, standpipe systems came into being as the result of ordinances passed when increasing heights and floor areas of new buildings began to pose severe firefighting problems. The time consumed in advancing hose lines several stories up to a fire floor by stairway, ladder, or hoisting with rope often was sufficient for a fire to get out of control. Some buildings were so tall that they did not allow any hose-line operations at all. The fire service urged that large structures be required by law to contain standpipe systems to eliminate the loss of time between their arrival at the fire ground and initiation of the fire attack.

The first laws, passed in major cities, required standpipes in all buildings more than 75 ft high. This is probably because at the time these laws were passed, the 75-ft aerial ladder was normally the longest ladder available; however, many thousands of buildings constructed under the old laws lack standpipes if they are more than three stories but fewer than 75 ft high. Fortunately, many cities, after finally passing or modernizing standpipe ordinances, have compelled the owners to install standpipe systems in new buildings and retrofit older ones.

NFPA 14, *Standard for the Installation of Standpipe and Hose Systems*, generally is referenced by local, state, and model building codes and insurance standards when designing a system. Other codes and standards adopted by the authority having jurisdiction also may influence system design.

Classes of Systems

Standpipe systems are designated class I, class II, and class III according to their intended use. The system could be designed for fire department use, first-aid firefighting, or both.

Class I   Class I systems provide 2½-inch hose connections at designated locations in the building for full-scale firefighting. These systems generally are intended for use by the fire department rather than fire brigades and building occupants.

Key Points

Standpipe systems are designated class I, class II, and class III according to their intended use.

Class II   Class II systems provide 1½-inch hose connections at designated locations in the building for first-aid firefighting. These systems generally are intended for use by fire brigades and building occupants before the fire department arrives. This type of system usually is supplied with hose, a hose rack, and cabinet and a nozzle installed at each hose connection.

Class III   Class III systems are a combination of class I and class II systems. They can be used for full-scale and first-aid firefighting. Class I and class II hose connections are provided with this system.

Types of Systems

Standpipe systems are also classified by "types" depending on whether the piping is filled with water or whether the piping is dry. In addition, it is dependent on whether the water supply for firefighting will be automatically, semiautomatically, or manually available.

Beginning with the 1993 edition of NFPA 14, standpipe system types were completely redefined. The result was the creation of five categories:

- Automatic wet
- Automatic dry
- Semiautomatic dry
- Manual dry
- Manual wet

Automatic Wet   Automatic-wet systems have piping that is filled with water at all times and have an automatically available

supply capable of supplying the water demand necessary for firefighting.

Automatic Dry   Automatic-dry systems having piping that normally is filled with pressurized air. These systems are arranged, through the use of a device such as a dry-pipe valve, to admit water automatically into system piping when a hose valve is opened. They are connected to an automatically available water supply that is capable of supplying the water demand necessary for firefighting.

Semiautomatic Dry   Semiautomatic-dry systems have piping that normally is filled with air that may or may not be pressurized. These systems are arranged through the use of devices such as a deluge valve to admit water into system piping when a remote actuation device located at a hose station, such as a pull station, is operated. They also have a preconnected water supply that is capable of supplying the water demand necessary for firefighting.

Manual Dry   Manual-dry systems have piping that normally is filled with air; these systems do not have a preconnected water supply. A fire department connection (FDC) must be used to maintain supply water for firefighting.

Manual Wet   Manual-wet systems have piping that normally is filled with water for the purpose of allowing leaks to be detected. The water supply for these systems typically is provided by a small connection to domestic water piping, and it is not capable of supplying firefighting water demands. An FDC must be used to supply water manually for firefighting.

In summary, wet systems have water-filled piping, and dry system do not. Automatic systems provide water supply for firefighting by simply opening a hose valve. Semiautomatic systems are connected to a water supply for firefighting but require activation of a device at a hose valve in addition to opening the valve to get water. Manual systems do not have a preconnected water supply for firefighting, and these systems must be supplied manually by connecting hoses from a pumper truck to an FDC.

Dry Systems

A dry system is simply a vertical pipe, or riser, running through or along the outside of the building. An interior dry system has at least one outlet on each floor inside the building; the fire department intake is located outside the building. An exterior dry system usually runs alongside the fire escape **Figure 10-1**.

Key Points

A dry standpipe system is a vertical pipe, or riser, running through or along the outside of the building. They are not equipped with interior lines for occupant use; only fire department pumpers can feed them.

There is one outlet at each landing, with the inlet placed at the bottom of the pipe, a few feet above the ground. Usually, such an exterior setup is the result of laws that are newer than the building. Only fire department pumpers can feed dry systems; thus, dry systems are not equipped with interior lines for use by occupants.

Dry systems may be found in unheated buildings, in buildings in cities that do not require wet systems, and in buildings built before retroactive standpipe laws were passed. Where laws permit, dry systems are preferred in some heated occupancies to prevent expensive water damage due to tampering by tenants.

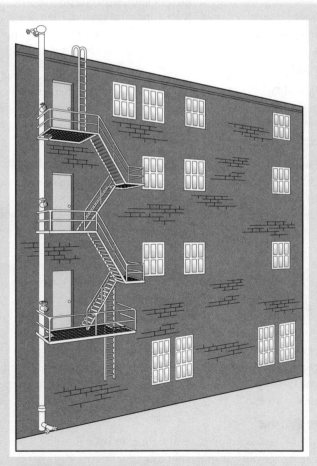

Figure 10-1 Dry standpipe systems may be found in tall or wide-area structures and can be located inside or outside the building.

Key Points

Dry standpipe systems are not equipped with interior lines for occupant use; they can be fed by fire department pumpers only.

Key Points

A building that is spread over a wide area may be equipped with two or more dry systems, which may be completely separate. Engine company personnel must be familiar with each separate system.

A building that is spread over a wide area may be equipped with two or more dry systems, each protecting a portion of the building. These systems might be completely separate, with each riser being supplied only by its own intake. It is important that engine company personnel become familiar with such separate systems during preincident planning activities or inspections of the property. Particular attention should be given to the location of the intake that feeds each riser. Obviously, the effect of pumping water into the wrong intake of a separate multiple-riser system would be disastrous.

Newer fire codes require that multiple-riser dry systems be interconnected so that water pumped into any intake feeds all the risers. One problem encountered with such systems is the time that it takes to drive the air out of a large system. This leads to a lag between the time when water first is pumped into the intake and the time when effective hose streams can be developed. Many departments have been caught off guard by this time lag, and the results usually are serious losses and unnecessary injuries to personnel. These systems should be time checked so that fire companies will know what to expect when using them.

Wet Systems

An automatic wet standpipe system is connected to a water source and contains water at all times **Figure 10-2**. In this system, the water must be under enough pressure to allow fire attack without aid from fire department pumpers. Meeting these residual pressure requirements assures sufficient water flow for normal initial attack operations.

NFPA 14 requires that a water supply for class I and class III systems be able to deliver a residual pressure of 100 psi at the

Key Points

An automatic wet standpipe system is connected to a water source and contains water at all times. The water must be under enough pressure to allow fire attack without aid from fire department pumpers.

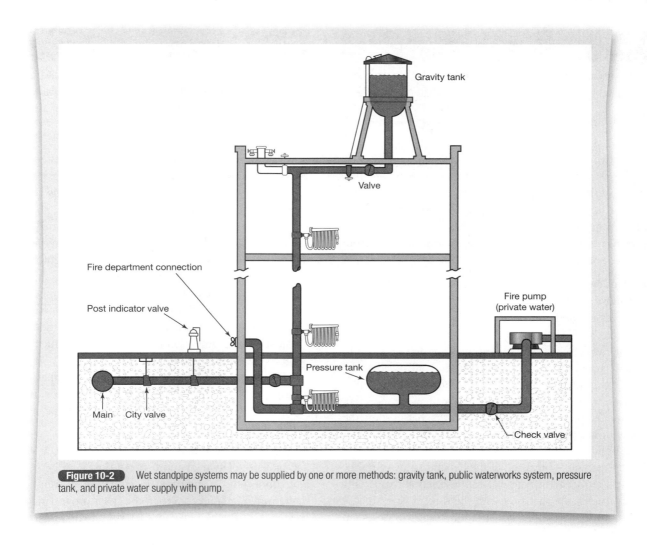

Figure 10-2 Wet standpipe systems may be supplied by one or more methods: gravity tank, public waterworks system, pressure tank, and private water supply with pump.

outlet of the top-most hose connection on each standpipe. This pressure must be available while flowing 250 gpm from each of the two topmost hose connections of the hydraulically most remote standpipe, plus 250 gpm from each additional standpipe, up to the required maximum. NFPA 14 requires that a water supply for a class II system be capable of delivering 100 gpm for 30 minutes. It also must be strong enough to maintain a residual pressure of 65 psi at the outlet for the hydraulically most remote hose connection with 100 gpm flowing.

NFPA 14 was modified in the 1993 edition to increase the minimum outlet pressure for class I and class III outlets from 65 psi to 100 psi because of questions raised regarding the adequacy of a 65-psi minimum design pressure on automatic and semiautomatic standpipes.

Water Supply to Wet Standpipe Systems

A fire department with a reliable and adequate source of water that is accessible nearby may supply manual standpipe systems. Automatic and semiautomatic standpipe systems require a minimum of one preconnected water supply that is capable of supply-

ing the standpipe system's hydraulic demand for the minimum required duration. In addition to the primary water supply on an automatic or semiautomatic standpipe system, one or more FDCs with a reliable water source accessible nearby are required.

For high-rise buildings, two remotely located FDCs are required for each zone within the pumping range of fire apparatus in addition to the automatic water supply. Two FDCs reduce the possibility of all supply hose being cut by falling glass, thus interrupting the secondary water supply during a fire. The sources of water most often used include public waterworks systems, gravity tanks, pressure tanks, and fire pumps connected to a reliable fixed water source, such as a pond.

Key Points

The most often used sources for water supply include public waterworks systems, gravity tanks, pressure tanks, and fire pumps connected to a reliable fixed water source.

Key Points

When two or more sources feed a wet standpipe system, the source providing the highest pressure will be the one that provides the water.

When two or more sources feed a wet standpipe system, the source providing the highest pressure will be the one that provides the water. If the supply in use fails, such as the building's fire pump or a fire department pumper, the remaining source with the most pressure cuts in and supplies the system. A check valve arrangement keeps possibly contaminated water from entering the potable water system.

Public Waterworks　When available, a public waterworks system almost always is used as the primary source of water. Other sources are used only as necessary because of extenuating circumstances, such as a broken or frozen hydrant or water main. If the public waterworks system supplies enough pressure to satisfy code requirements, no other source will need to be used. The exception is in industrial, warehousing, and similar occupancies, where additional water sources may be required. A wet system, supplied by public waterworks, has become the most common arrangement in medium- and high-rise apartment and office buildings.

Fire fighters must be aware of the dependability of public waterworks supplies in their areas of responsibility and especially of changes in pressure that vary with the time of year, time of day, and so on. For example, will a public waterworks system deliver as much water at 1600 hours on August 15 as it will at the same time on December 15? Will the demand on the water system during the day have the same effect as in the middle of the night? It is vital that the answers to such questions be determined and that this information be used in a fire situation.

Other Sources　If the public waterworks system is insufficient or unavailable for a wet standpipe system, other sources must be used. Pressure pumps, for example, are used to boost the pressure in the primary supply, which usually is the public waterworks system but could be a water tank. Gravity tank systems are another alternative; they are mounted on the ground when supplying one- and two-story buildings or are mounted

Key Points

Fire fighters must be aware of the dependability of public waterworks supplies in their areas of responsibility and especially of changes in pressure that vary with the time of year, time of day, and so on.

Key Points

If the public waterworks system is insufficient or unavailable for a wet standpipe system, other sources must be used. Other sources include pressure pumps and gravity tanks.

above ground in taller structures. They may hold as much as 100,000 gallons of water. Gravity supplies the pressure required in the standpipe system. A pump is used to fill the tank but not to aid in developing pressure.

Pressure tanks generally are found only on smaller systems, where they supply house lines with water for initial attack by the occupants. They also may be used to augment another source in case of failure in the main supply. They rarely exceed 3,000 gallons in capacity and can be located anywhere in the water supply system. The pressure tank differs from the gravity tank in that the top third of the tank contains air under pressure. This assures sufficient water pressure while the tank is supplying water to the standpipe system.

Remember that a water tank has a given capacity. After it is emptied of water, it is useless as a source. For this reason, pumpers should be used to supply any standpipe system, wet or dry, after firefighting operations begin.

Multiple Wet Systems　As with dry systems, a wet system may consist of two or more risers that may be separate or interconnected. In any case, the risers are all supplied by the same source or sources. When the risers are separate, there is one fire department intake for each riser; each intake serves one riser and one riser only. When the risers are interconnected, the system must have two or more intakes　**Figure 10-3**　. Again, it is important that fire fighters be aware of the type of system with which they are working.

In the case of a separate-riser system, the multiple intakes allow water flows to various areas of a building without the need

Key Points

After a water tank is empty, it is a useless source; therefore, a pumper should be used to supply any standpipe system, wet or dry, after firefighting operations begin.

Key Points

As with dry systems, a wet system may consist of two or more risers that may be separate or interconnected. Fire fighters must be aware of the type of system with which they are working.

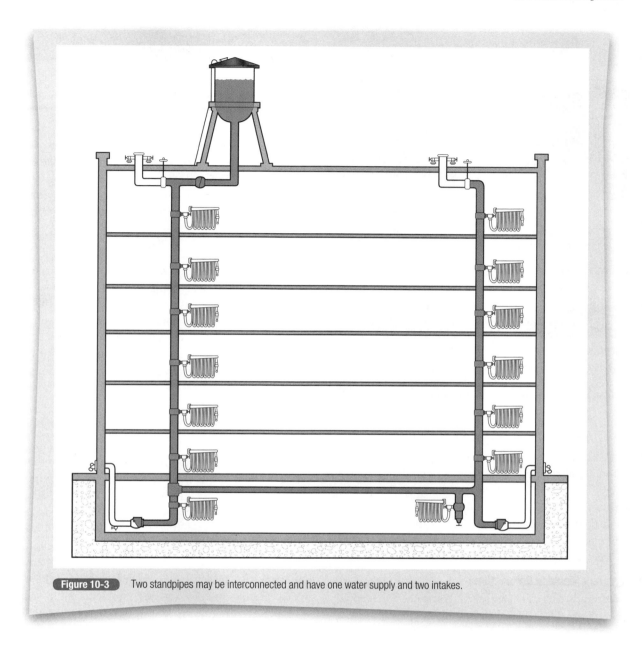

Figure 10-3   Two standpipes may be interconnected and have one water supply and two intakes.

for long hose-stretching operations. In addition, if one riser is out of service, the second can be used for fire attack although a long stretch of hose may be required on the fire floor. If a 1¾-inch hose is used for the firefighting operation, a 2½-inch hose line should be used to bring the water close to the fire area. It then should be wyed off to 1¾-inch hose lines, as a long stretch of hose of these smaller sized hose lines would create excessive friction loss and reduce the water delivery. It may be advantageous to use the 2½-inch hand line for fire attack and forgo the use of the smaller hose line.

Valves   A wet standpipe system must have some type of cutoff or control valve between the system and the water main or other water source in order for the system to be

shut down for repairs and maintenance. Postindicator valves usually are found on industrial and warehouse properties; gate valves or outside stem-and-yoke valves usually are installed in apartment buildings, office buildings, and stores.

Larger systems can be divided into controlled units or areas, permitting only part of the protection to be cut off to allow repairs. In such cases, there may be several valves. Before attempting to use a standpipe system, fire fighters should check to make sure the postindicator valves are in the "open" position. Postindicator valves may be found on wet standpipe systems. Two popular types are the window type and the butterfly type **Figure 10-4** .

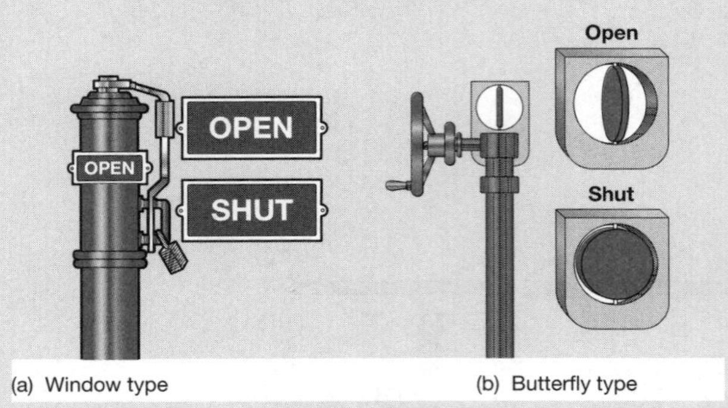

(a) Window type   (b) Butterfly type

Figure 10-4   The (a) window type and (b) butterfly type are two commonly used postindicator valves.

When such a valve is not placed on the outside of the building, a gate valve, or outside stem-and-yoke valve, might be located on the intake pipe in the basement or just inside the first floor if the building has no basement. These valves normally are placed close to the outside wall of the building, with many variations in their exact locations. Whatever the location, fire department members should know the locations of these valves during an incident.

Fire Department Siamese Connections

FDCs are required on all class I and class III systems. On manual standpipe systems, FDCs serve as the only water supply. On automatic and semiautomatic standpipe systems, they serve as an auxiliary water supply.

As a rule, most FDCs are provided with one 2½-inch inlet for each 250 gpm of design flow rate. Some fire departments require large-diameter inlets that are designed to connect to large diameter hose. It is preferable to use 4-inch or larger piping for connecting the FDC to the system mains. Water pumped

into the siamese connection always will reach the riser outlets, as there are no valves in the riser between the siamese and the outlets. **Figure 10-5** shows several siamese-type connections: wall-mounted siamese, free-standing siamese, free-standing multiple siamese, and wall-mounted concealed siamese.

Intake Location   Fire fighters should be familiar with the locations of these intakes, especially if local codes do not specify their position. Signs should be provided to indicate whether a connection serves a standpipe system, a sprinkler system, or both. In addition, the signage should indicate the area being serviced by the connection. The connection and the signage should be accessible and unobstructed.

If the intakes are not placed on the wall, but rather on a pipe that stands away from the wall, the designations appear on top of the pipe, just behind the point at which the two intakes wye off. Some cities require a color code to indicate the type of system fed by each intake, but local requirements vary.

Hooking Up to the Siamese Connection   Remember that problems can arise with any standpipe water supply system. Fire fighters should check the system to ensure that all components are functioning properly. This will involve checking that valves are open and the fire pump, if so equipped, is operating. The pumper should be hooked up to the standpipe system to ensure a water source if the building's system is out of order or ineffective. There is no sense in taking chances when the building's system might not be operative. Check valves could be frozen into a closed position. Water pumps could fail, or other problems might develop.

Before hooking up to the standpipe system, the FDC should be checked for debris. These connections are a target for the collection of rocks, cans, bottles, and other trash. At least two supply lines should be connected into the intake siamese **Figure 10-6**. The first line should be hooked to the left intake and charged to get water into the system quickly. The second line then should be connected to the right intake. This makes it easier to line up a hookup; it also aids in identification of the supply lines if trouble should develop. The clapper valves in the

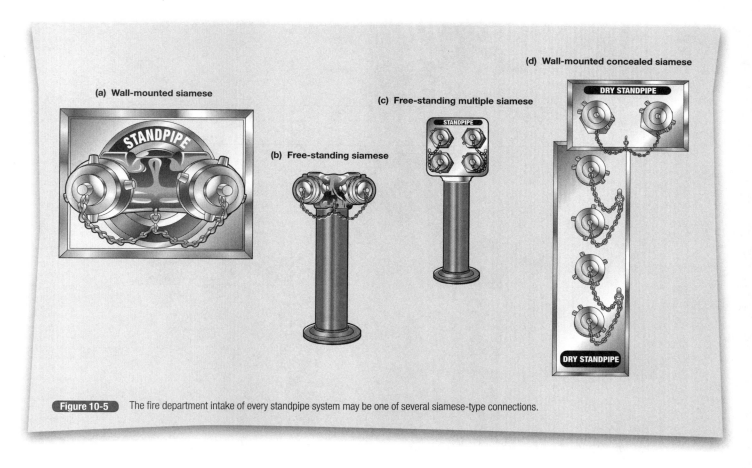

(d) Wall-mounted concealed siamese

(a) Wall-mounted siamese

(c) Free-standing multiple siamese

(b) Free-standing siamese

Figure 10-5 The fire department intake of every standpipe system may be one of several siamese-type connections.

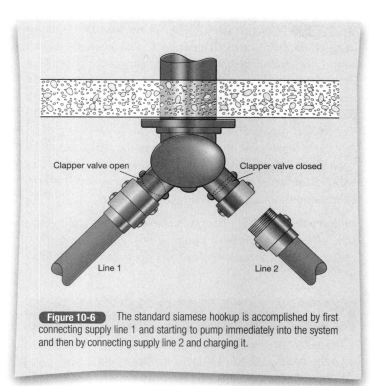

Figure 10-6 The standard siamese hookup is accomplished by first connecting supply line 1 and starting to pump immediately into the system and then by connecting supply line 2 and charging it.

siamese connection will prevent water from discharging out of the right-hand connection after the left side is charged.

As noted, many departments are also using large diameter hose to supply standpipe systems. The standpipe connection itself may be equipped with a single adapter for large-diameter hose. It is extremely important that the hookup be accomplished quickly so that fire fighters with hose lines will have water under effective pressure if and when they need it. If a dry standpipe system is in use, the pumper will hook up to the system and supply the source of water.

A wet system will draw water from its source until a fire department pumper increases its pressure to be greater than the source pressure. When this occurs, the building system's check valves will close to keep the pumper water from flowing to the source, and the pumper will supply the system. If the pumper pressure is reduced for any reason, the normal source again supplies the system.

Key Points

It is extremely important that the siamese hookup be accomplished quickly so that fire fighters with hose lines will have water under effective pressure if and when they need it.

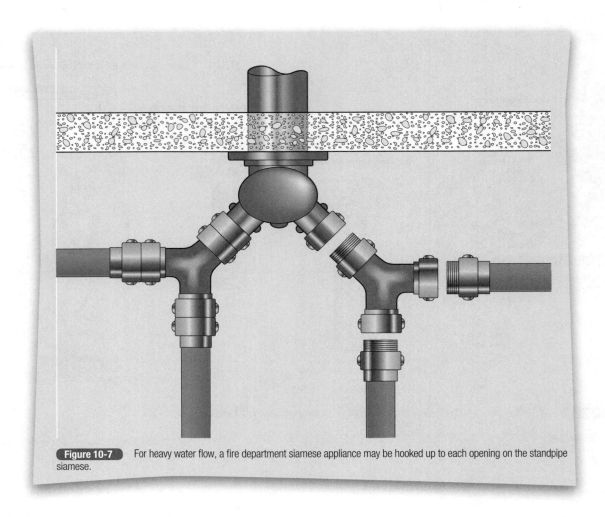

Figure 10-7   For heavy water flow, a fire department siamese appliance may be hooked up to each opening on the standpipe siamese.

When the fire is severe, a fire department siamese appliance can be placed on a standpipe siamese intake, and additional supply lines can be hooked into the system **Figure 10-7**. Many fire departments mandate that standpipe connections be equipped with a 4-inch hose connection instead of multiple 2½-inch hose connections. Remember, the minimum size of the standpipe is 4 inch, and thus, its water-carrying capability is great.

Damaged Siamese Connection   At many building sites, siamese connections are not well maintained, and they could be damaged. If a supply line cannot be connected to the siamese, water can be supplied to the system through the hose line outlet on the first floor. First, the house line and any pressure reducing fittings must be removed, and suitable fittings must be installed

Figure 10-8. Whenever possible, a 2½-inch siamese should be installed on the outlet with a double female adapter, and two supply lines should be used to deliver water to the outlet from the pumper. If the location of the outlet prevents the use of a siamese, a single supply line can be hooked up to the outlet with the double female adapter. The use of large-diameter hose also should be considered.

For a wet system, the supply line to the outlet should be hooked up and charged before the outlet valve is opened. This will prevent water in the system from running back into the supply line and possibly impeding the operation.

Pumper Positioning   The best arrangement for supplying a standpipe system is to have a pumper positioned within 100 ft of the standpipe siamese. As in other operations, this is more

Key Points

Siamese connections often are not well maintained and could be damaged. If a supply line cannot be connected to the siamese, water can be supplied to the system through the hoseline outlet on the first floor.

Key Points

The best arrangement for supplying a standpipe system is to have a pumper positioned within 100 feet of the standpipe siamese.

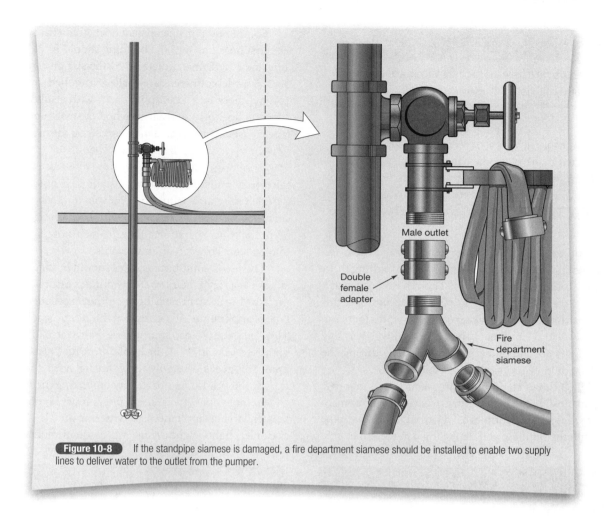

Figure 10-8 If the standpipe siamese is damaged, a fire department siamese should be installed to enable two supply lines to deliver water to the outlet from the pumper.

efficient than pumping through several hundred feet of hose. Hose may be laid from an adequate hydrant to the pumper at the building or from a pumper at the hydrant or other water source to the pumper at the building. This keeps control of the supply lines to the standpipe system close to the system itself. Also, communications between the pumper and fire fighters on the hose lines inside the building are simplified.

If the standpipe system is equipped with a fire pump, the operating pressure of this pump should be checked to help determine the proper operating pressure of the pumper if it is to be used.

Fire Attack from Standpipe Systems

A fire that requires the use of a standpipe system for water supply is most likely on a floor higher or lower than the ground floor and therefore some distance from the fire apparatus. Before they can attack the fire, fire fighters will have to carry all their equipment into the building, get it up to or near the fire floor, and set up to use the standpipe outlets. As fire fighters proceed up the stairs, they should check to ensure that all hose outlet valves on the lower floors are closed. Open valves will take water away from the area of the fire.

Equipment

Equipment necessary to establish effective operations quickly and the safety precautions related to standpipe operation are covered in this section.

House Lines Fire department personnel never should rely on house lines that are intended for use by building occupants. These hose lines often are poorly maintained and rarely are tested. As a result, hose often is found to be partially charged, snarled, and/or in disarray in the cabinet or hose rack. Hose also could be missing from the standpipe location. Unlined hose, if

Key Points

Fire department personnel never should rely on house lines that are intended for use by building occupants. These hose lines often are poorly maintained and they are tested infrequently.

still in use, may be rotted. Valves, never used or tested, frequently cannot be opened by hand, or the handwheel may be missing. Engine company crews should carry their own hose lines into a fire building and be cautioned never to attempt to use building's house lines. **Figure 10-9** illustrates hose lines in poor and good condition.

Hose and Appliances Fire departments should have the necessary tools and equipment and a way of transporting them expeditiously to the desired location. There have been many innovative and creative ideas put in place to accomplish this task. Each department should have a system that works for them. Hose and all tools and equipment should be lightweight. Only equipment that is absolutely necessary, especially during the initial attack, should be brought into the building.

Figure 10-10 shows standpipe equipment stored in a bag. This arrangement allows two fire fighters to share the equipment load and provides an efficient means of carrying the standpipe equipment into the building and deploying it at the outlet.

Some items remain constant no matter what system is used. The first hose line into the building should be at least 150 feet long and of a diameter dictated by the size and intensity of the fire. Many departments carry flaked, prerolled, or donut-rolled 1¾-inch hose in a "standpipe pack" with a solid-bore nozzle, adapters, spanners, and a handwheel or small pipe wrench. The latter is used to open the standpipe valve if the handle is missing or if the valve cannot be opened by hand.

Hose lines that are 1¾ inches are effective on most fires in the typical building containing a standpipe system; however, a 2½-inch hose also should be available for back up in case the smaller hose lines are inadequate for controlling the fire. As in all firefighting operations, the 2½-inch hose should be used immediately if the situation warrants.

The use of solid-bore nozzles should be considered for fire attack because of their lower operating pressures. Modern low-pressure break-apart nozzles also are an excellent choice for this type of application. An example of this nozzle assembly includes a low-pressure (75 psi) spray nozzle with a rated discharge of 150 gpm coupled with a $^{15}/_{16}$-inch solid-bore tip (50 psi) rated at 180 gpm. This nozzle assembly allows fire fighters a choice of a spray nozzle or a solid-bore nozzle for combatting the fire.

In large open areas, fire fighters must be prepared to use master stream appliances for interior attack. The single or two-inlet, lightweight deluge set, made of modern alloys and

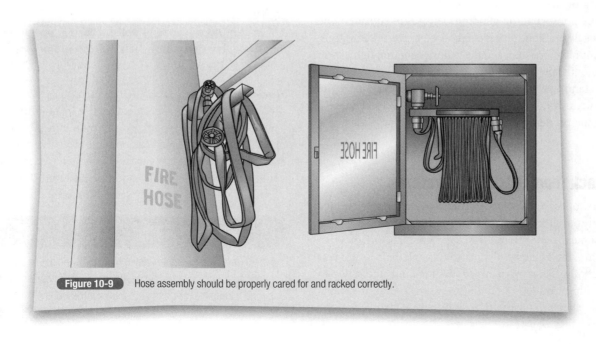

Figure 10-9 Hose assembly should be properly cared for and racked correctly.

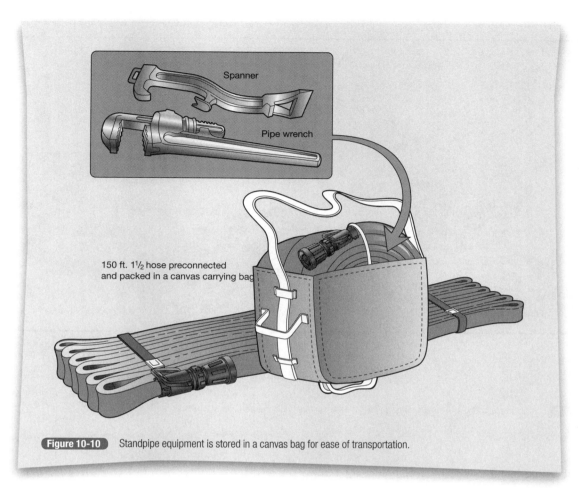

Spanner

Pipe wrench

150 ft. 1½ hose preconnected
and packed in a canvas carrying bag

Figure 10-10 Standpipe equipment is stored in a canvas bag for ease of transportation.

equipped with a solid-bore tip, has proved most effective for such operations.

Tools   Forcible entry tools must be carried into the building to be used as needed by fire fighters. Conducting a primary search, attacking the fire, checking for fire extension, and performing other tasks may require fire fighters to force their way into locked apartments, offices, or other secured areas of the building. Other tools and equipment may be needed inside the building to assist in firefighting operations. Smoke ejectors and positive pressure fans are helpful in setting up horizontal and vertical ventilation when they are needed. Portable lighting may be necessary to illuminate smoke-filled or windowless areas of the building.

An equipment pool should be set up usually two or more floors below the fire, depending on conditions on the fire floor or other ongoing activities within the building **Figure 10-11**. Additional equipment necessary to support the operation should be brought to this interior staging or resource area.

Key Points

Fire fighters must carry forcible entry tools into the building to be used as needed. Smoke ejectors, positive pressure fans, and portable lighting also are useful.

Key Points

All of the equipment necessary to support the firefighting operation should be brought to an interior staging or resource area, also called an equipment pool, usually located two or more floors below the fire.

Entering the Building

Safety   Buildings with standpipe systems generally are large and often are high rises, requiring the efforts of many fire fighters. Every person engaged in the firefighting or rescue operations must be in full personal protective equipment, including self-contained breathing apparatus. Moreover, no one is to be assigned to work in the building alone; all assignments should be given to teams of at least two fire fighters. Because of the fire-resistant construction of many standpiped buildings, the heat, smoke, and gases are difficult to remove. Long corridors filled with smoke cut visibility to a minimum. Fire fighters working together can keep tabs on each other by touch, voice, and radio

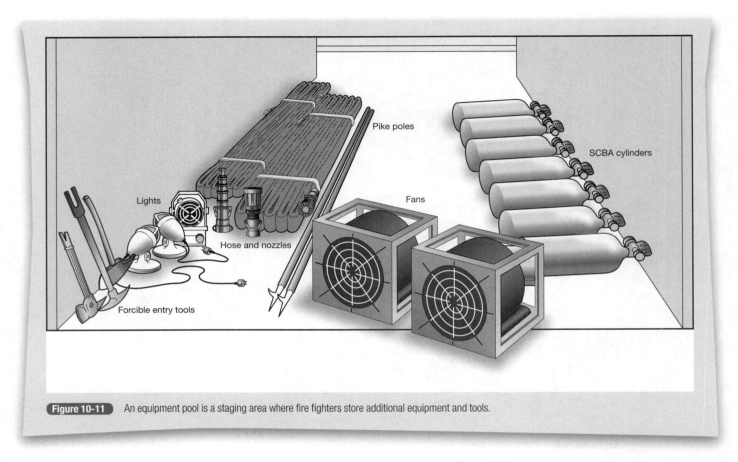

Figure 10-11 An equipment pool is a staging area where fire fighters store additional equipment and tools.

communications and contact with the hose line or rope lines stretched for that purpose. A fire fighter working alone is in great danger of being injured or lost. Communications by hand-held radio or by telephone from floor to floor should be established so that personnel can be moved quickly in case of an emergency. A personnel accountability system must be in place at every incident. This is a part of the incident management system and cannot be overlooked.

Checking Above the Fire Fire fighters should be assigned to check the floor above the fire, with hose lines if possible but without them if necessary. All floors, especially those immediately above the fire, are considered exposures. Hose lines need to get to floors above the fire to support search and rescue teams conducting the primary search, forcible entry, and evacuation efforts. Fire fighters need to check for fire extension or engage in battling the fire.

Fire tends to extend vertically because of lapping, or leap-frogging, from window to window **Figure 10-12**. Lapping refers to fire that jumps from floor to floor through exterior windows. This especially is true of fire-resistant buildings, a type of construction that tends to force the heat out at the windows, the weakest part of the structure.

Key Points

Every person engaged in the firefighting or rescue operations must be in full personal protective equipment, including self-contained breathing apparatus.

Key Points

No one is to be assigned to work in a building alone; all assignments should be given to teams of at least two fire fighters.

Key Points

A personnel accountability system must be in place at every incident.

Key Points

Fire fighters should check the floor above and below the fire for fire extension, with hose lines if possible but without them if necessary.

Figure 10-12 Personnel should be assigned to check above the fire floor for fire spreading from window to window.

Other avenues of travel include stairways and elevator shafts as well as utility shafts or pipe chases. A lack of a hose line or at least a check in the area over the fire may allow the fire to extend unnecessarily.

Checking Below the Fire   Floors below the fire also must be checked for fire extension. If fire is observed below the fire floor, hose lines will need to be stretched to these areas for fire control. In addition to fire extending to floors below, property conservation should be considered early in the incident. High-rise buildings, depending on their occupancies, may contain valuable contents. Protecting these valuables, especially from water damage, will reduce loss considerably. Property conservation on floors below the fire usually is labor intensive and because of other tasks being conducted on the fire ground may be overlooked in the initial stages of the fire.

Use of Elevators   Fire departments must have written standard operating guidelines that address the use of elevators during a fire incident. All members must understand and follow the elevator protocol. The incident commander must conduct a risk versus benefit analysis before allowing fire fighters to use an elevator in a fire situation.

Fire fighters should avoid using the elevators unless they can be operated safely and under supervision. In high-rise fire situations, the elevators and stairways are managed by lobby control. Elevators should not be used for fires on lower floors of a building. Fire fighters should use the stairways to gain entry to the fire floor.

An elevator never should be taken to the fire floor or any floor above the fire. Fire department personnel should not be allowed to break this rule. Standard operating guidelines should state that fire fighters using an elevator must stop two floors or more below the fire and then use the stairway to the fire floor. If the elevator is taken to the fire floor, the fire fighters in the elevator could be exposed to flames, excessive heat, or toxic gases when the door opens.

When the location of the fire is difficult to judge or the fire is reported to be on a certain floor, fire fighters can expect to be ordered off the elevator two to four floors below the suspected fire floor. This should be done even if smoke, heat, or flame-detecting devices are used to determine the fire floor.

Fire fighters must be in full protective clothing, including self-contained breathing apparatus. Fire fighters must bring forcible entry tools with them when using an elevator. Forcible entry tools will allow them to open or close the car or floor door as needed in an emergency or to force their way physically out of a stalled elevator.

When an elevator is available, it also can be used as an equipment lift, after personnel are in position. Items required for fire attack may be sent up to the interior staging area by fire fighters on the ground floor.

Fire fighters should become completely familiar with the elevators in buildings in which they may have to be used. Representatives of elevator companies should be consulted, and training should be conducted for accurate information on the proper operation of an elevator's system during a fire incident. Guidelines must be in place to ensure the safe use of elevators during an emergency.

If the elevators have lobby controls, their locations should be known. Fire fighters should know how to obtain and keep control of elevators. Modern codes require a fire service control in each car as well as in the lobby. This gives fire fighters control of the car and cuts off the response to call buttons on each floor.

Beginning Attack Operations

Fire fighters, having taken an elevator to a point below the fire, will advance by stairway to the fire floor. In many cases, the first hose line will be connected to a standpipe outlet in the enclosed stairwell. A hose connection located one floor below the fire floor or on an intermediate landing should be used over the connection on the fire floor so that there will be room to spread out the hose line before it is charged. This should provide for less congestion in the stairway of the fire floor as the attack gets underway. If 1¾-inch hose lines are to be used for initial attack, a wye can be placed on the outlet or attached to a short length of 2½-inch hose, and two attack lines can be hooked to it.

No matter how good the house lines look, they should always be removed from the standpipe outlet, and fire department hose should be hooked up for firefighting **Figure 10-13** . Under no circumstances should fire fighters use the hose found on the outlet.

When a hose line is connected to a standpipe outlet in a stairwell on or below the fire floor, the excess hose should be pulled up the stairway toward the next floor before it is charged. The hose will come down the stairs easily as the advance is made. On the other hand, if the hose is thrown down a stairway, it must be worked, or pulled up, after it is charged. If part of the corridor is involved, the hose line should be charged before it is advanced from the stairwell to the corridor **Figure 10-14** .

If smoke or fire is detected on the fire floor, hose lines should be stretched from a standpipe connection in the stairway and not from a standpipe connection in a corridor. The theory here is that a charged hose line should be in place before entering the corridor or hallway.

Most of the hose line should be taken up the stairs, so it can be advanced more easily through the corridor of the fire floor. If required, additional lines can be taken up the stairs from still

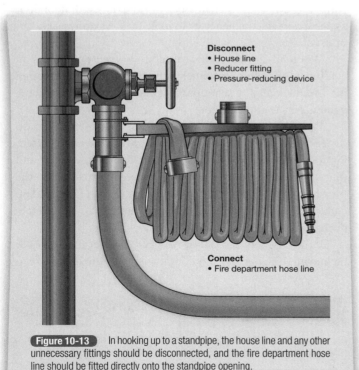

Figure 10-13   In hooking up to a standpipe, the house line and any other unnecessary fittings should be disconnected, and the fire department hose line should be fitted directly onto the standpipe opening.

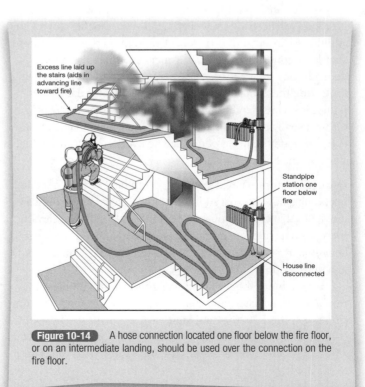

Figure 10-14   A hose connection located one floor below the fire floor, or on an intermediate landing, should be used over the connection on the fire floor.

Key Points

If smoke or fire is detected on the fire floor, hose lines should be stretched from a standpipe connection in the stairway and not from a standpipe connection in a corridor.

Key Points

When fire fighters are using the stairwell to advance hose lines, they must be careful not to impede the progress of evacuees.

lower floors. This might also be necessary if the floor below the fire is untenable **Figure 10-15**.

If the fire is located some distance down the corridor from the stairway, the initial hookup can be made on the fire floor; however, this should not be attempted unless fire fighters are certain that the fire is confined to an area off the corridor or at least is some distance from their point of entry to the fire floor.

It is obvious that the stairwell nearest the fire area is important for advancing hose lines to the fire floor. It is just as important to occupants of the building who may be using it for evacuation. Fire fighters must be careful not to impede their progress and not to allow great volumes of smoke to get into the stairway. If another stairway, further from the fire, is available, evacuees should be directed to it. Sometimes, if conditions are favorable, as is often the case in fire-resistant structures, the best thing is to convince occupants to stay in their offices or apartments.

In many cases, the first hose lines advanced from the floor or floors below the fire will control the fire enough to permit a standpipe

outlet on the fire floor to be used safely. It then can be used either for backup lines or for additional attack lines; however, fire that has moved into a corridor creates intense heat all along the hallway. Fire fighters always should be prepared to fight their way into the corridor from the stairwell; when a long corridor is completely involved, streams should be directed as far into the corridor as possible.

After fire fighters are in the corridor of the fire floor, the fire is attacked in the usual way. Search and rescue operations should begin as soon as possible, and areas of the building should be checked for fire spread.

When fire has gained substantial headway in a standpiped structure, 2½-inch hose lines should be used for the initial attack. This especially is true of fire in a building not divided into apartments or offices. Where it is known that large open areas are involved, solid-stream nozzles should be used, and master stream appliances should be taken into the building.

A master stream appliance may be supplied from one or more standpipe outlets as required, with 2½-inch or large-diameter hose **Figure 10-16**. For efficient operation, lightweight hose with lightweight couplings, as recommended for use with other standpipe equipment, should be used with the master stream appliance.

Fire-Resistant Structures

Most fire fighters will encounter standpipe systems in low-, medium-, and highrise-structures of so-called fire-resistant construction. They must not become overconfident or complacent in handling fires in these buildings simply because the type of construction and the presence of standpipes seem to indicate an easily controlled fire. Serious fires, causing death and injury to fire fighters and occupants, have occurred in such buildings in the United States and elsewhere.

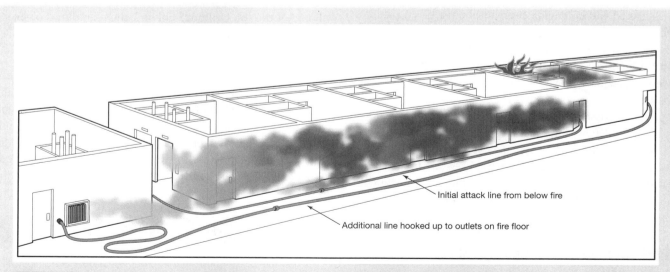

Figure 10-15 When the standpipe outlets are in a corridor, initial attack should start using outlets on the floor below the fire. Then it may be possible to use the outlets on the fire floor.

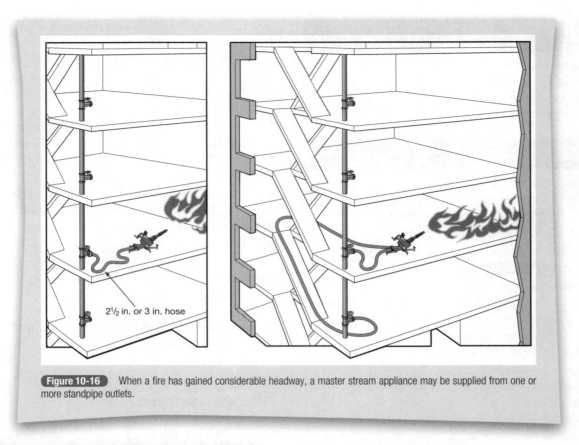

Figure 10-16 When a fire has gained considerable headway, a master stream appliance may be supplied from one or more standpipe outlets.

2½ in. or 3 in. hose

Key Points

Most fire fighters will encounter standpipe systems in low-, medium-, and high-rise structures of fire-resistant construction. They must not become overconfident or complacent in handling fires in these buildings.

As stated earlier, engine companies must be prepared to attack fires in these structures with streams ranging from 1¾-inch lines to master stream appliances. Many departments have only 1¾-inch line setups for immediate use with standpipes; however, these have proved ineffective for fire control on many occasions, not only in buildings with large open areas but also in apartment and office buildings where fire has gained control of a corridor and/or wind is fanning the fire.

This condition creates what some fire officials term a "blowtorch effect." In several such instances, fire fighters have been unable to advance from the stairway into the corridor with 2½-inch lines. The ineffectiveness of 1½- or 1¾-inch lines under these conditions is then obvious.

Other Uses for Standpipe Systems

Standpipe systems, although primarily designed for interior fire attack, can be used to attack fires in adjoining buildings and for exposure coverage.

Fire Attack in Adjoining Buildings

When fire has gained considerable headway in a building, fire fighters may be kept from the fire floor by the intense heat. Under these conditions, if adjoining buildings are close enough to the fire building, hose streams developed from standpipe systems in the nearby buildings might be used on the fire **Figure 10-17**. They can be directed across a court, across narrow streets and alleys, or from the roof of the adjoining building into higher floors of the fire building. Either 1-inch handlines or master stream appliances may be used for such operations.

Exposure Protection

Fire on the roof and in upper floors of closely set buildings present serious exposure problems. In some cases, aerial line apparatus can be used for fire attack and exposure protection; however, when the fire is above the reach of aerial devices, almost all operations must depend on the standpipe system of the fire building and adjoining structures for water. Pumpers must be set up to pump into the standpipe systems of both the fire building and exposed buildings. The standpipe system of the fire building

Key Points

When the fire is above the reach of aerial devices, almost all operations must depend on the standpipe system of the fire building and the adjoining structures for water.

Figure 10-17 Standpipe systems may be used for fire attack and exposure protection from adjoining buildings.

must be charged to make interior hose lines usable. Those of the exposed buildings must be charged to assure proper water supplies if and when they are needed. This is especially important when only public waterworks supply the systems because the public waterworks pressure could be reduced to the point where attack lines supplied by these standpipes would be ineffective.

Use of Water from Uninvolved Buildings

Water from a gravity tank atop an uninvolved building can be drawn off and used to supply pumpers **Figure 10-18**. This is usually a measure of last resort; it is used, for example, when a very large fire requires many pumpers to operate simultaneously so that the water main pressure may become dangerously low. To obtain the water, the pumper is parked as close to the ground-floor standpipe outlet as possible. The 2½-inch hose or larger supply line is hooked up between the outlet and the pumper intake. The outlet is opened, and the pumper takes water from the gravity tank. Depending on the standpipe system, it could be more efficient to use large-diameter hose to support the operation.

Automatic Sprinkler Systems

When properly designed, installed, maintained, and supported by the fire department, a sprinkler system can apply water directly to the fire in a more effective manner than can the fire department using manual fire-suppression methods. In such a system, water is piped from a source to the protected area. There, the piping is branched, usually at ceiling level, to cover the entire area. Small nozzles, called heads, are placed at intervals in the piping. The heads are sensitive to heat; even a fairly small fire will cause them to open and apply water to the area below. As long as enough pressure is available, each head will distribute water over an area of 100 ft^2 or more in volume sufficient to control or extinguish most fires.

Key Points

A properly installed and maintained sprinkler system is extremely effective in protecting a building and its contents from fire.

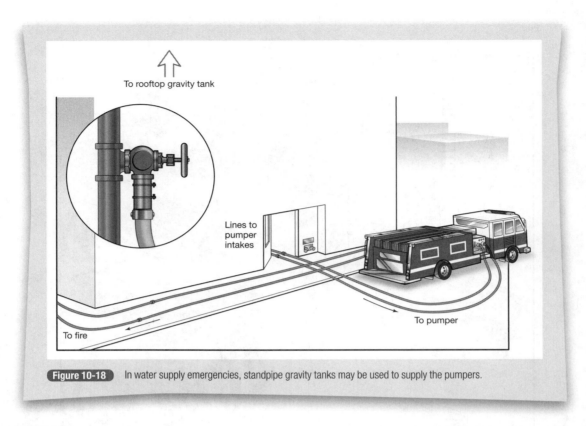

Figure 10-18 In water supply emergencies, standpipe gravity tanks may be used to supply the pumpers.

Types of Sprinkler Systems

There are four basic types of sprinkler systems:

Wet-pipe system	Preaction system
Dry-pipe system	Deluge system

Wet-Pipe System In the wet-pipe system, the piping always is filled completely with water from the source to the heads. When a sufficient amount of heat is generated from a fire, water is discharged immediately from an activated sprinkler or sprinklers. This is the simplest and one of the least expensive types of system, requiring little in the way of special equipment; however, it cannot be used in unheated buildings or for exterior protection in areas where the temperature might go below freezing. Nevertheless, it is the most common type of sprinkler system.

Dry-Pipe System The dry-pipe system was developed for use in unheated areas where the piping in a wet system would freeze. The system contains water only from the source to a control valve, known as the dry-pipe valve **Figure 10-19**. The piping from the valve to the heads contains compressed air. The air pressure, maintained by a compressor, holds the dry-pipe valve closed. Most dry-pipe valves act on a pressure differential principle, in which the surface area of the valve face on the air side is greater than the surface area on the water side. When a sprinkler head opens, air escapes from the lines. This reduces the pressure and trips the dry-pipe valve, which opens and allows water to flow into the sprinkler piping and then to the open head or heads. In unheated areas, the dry-pipe valve is enclosed in an insulated and heated closet or cabinet to protect it from freezing. The enclosure usually is small enough to allow protection with minimum heating and large enough to permit entry for maintenance.

Preaction System When a dry system is activated, there is a lag between the time the head opens and the time water is applied to the fire. The preaction system was devised to minimize this time lag by exhausting the compressed air before the head opens.

The preaction system is a dry-pipe system with the addition of air exhausters, if so equipped, controlled by a supplemental fire detection system. When the supplemental fire detection system

> ### Key Points
>
> In a wet-pipe sprinkler system, the piping always is filled completely with water from the source to the heads. When a sufficient amount of heat is generated from a fire, water is discharged immediately from an activated sprinkler or sprinklers.

> ### Key Points
>
> The dry-pipe system was developed for use in unheated areas where the piping in a wet system would freeze.

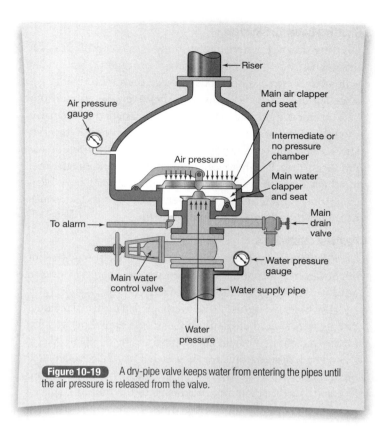

Riser

Air pressure gauge

Main air clapper and seat

Intermediate or no pressure chamber

Air pressure

Main water clapper and seat

To alarm

Main drain valve

Main water control valve

Water pressure gauge

Water supply pipe

Water pressure

Figure 10-19 A dry-pipe valve keeps water from entering the pipes until the air pressure is released from the valve.

is activated, the air exhausters, which are located at the dry-pipe valve, open and let the compressed air escape from the system. This allows water to flow into the sprinkler piping and to the heads and, in most cases, to be at the head when the temperature rises sufficiently to cause it to open. Some such systems have additional exhausters on the sprinkler piping to increase the speed of air release and water travel to the heads. Preaction systems can be used wherever dry-pipe systems are used.

Deluge System Designed to deliver large volumes of water quickly, the deluge system differs in the basic from the systems discussed previously. A deluge system contains water only from the source to a control valve. The piping from the valve to the heads contains only room air, and the heads are open at all times. Complete control of the system is provided at the valve. Normally, compressed air keeps the valve closed. If there is a fire, the valve is activated by a supplemental fire detection system. The valve then opens and allows water to flow through the piping and out of all of the heads simultaneously. This floods an entire area quickly, which is what the system is designed to do.

Deluge systems usually are installed in high-hazard locations such as aircraft hangars, chemical plants, laboratories and transformer rooms, or other areas where fire might spread rapidly. These systems must have good water supplies in order to function effectively.

Water Supply for Sprinkler Systems

According to NFPA 13E, *Recommended Practice for Fire Department Operations in Properties Protected by Sprinkler and Standpipe Systems*, when arriving at a property protected by an automatic sprinkler system, fire companies should take prompt actions to supply the system. Any of these systems can be supplied by a public water works system, gravity tanks, pressure tanks, and fire pumps connected to a reliable fixed water source. Every sprinkler system must be equipped with a fire department intake.

Following department standard operating guidelines and NFPA 13E, an engine company, on arrival at a fire in a sprinkler-protected structure, immediately should hook up to the siamese connection and prepare to pump into the system. Fire fighters should check the system, including the main valve(s), the sprinkler riser, and the fire pump, if so equipped, to ensure that all components are functioning properly. An actual check of the fire pump is needed. The incident commander needs to know whether the fire pump is functioning. If the pump is not functioning, it may be possible to have the pump manually started by building personnel. During an actual fire incident, a fire fighter must be assigned to the main valve and fire pump to ensure a continued operation of the system. If there is a confirmed fire, a pumper should be hooked up to the sprinkler system to ensure a water source if the building's system is out of order or ineffective. Given a continuous supply of water, the sprinkler system will fight the fire effectively; the job of the engine company is to assure that the water supply is adequate.

The pump operator should wait for orders to begin pumping. Department standard operating guidelines should dictate the hose layout and connection to the sprinkler connection as well as the initial pump pressure. Books and publications on supplying

Key Points

The preaction system is a dry-pipe system with the addition of air exhausters, if so equipped, that are controlled by a supplemental fire detection system.

Key Points

Deluge systems usually are installed in high-hazard locations, such as aircraft hangars, chemical plants, laboratories, and transformers. They must have a good water supply to function effectively.

Key Points

Every sprinkler system must be equipped with a fire department intake.

> **Key Points**
>
> An engine company, on arrival at a fire in a sprinkler-protected structure, immediately should hook up to the siamese connection and prepare to pump into the system.

> **Key Points**
>
> After pumping begins, the pump operator must monitor the discharge pressure carefully.

> **Key Points**
>
> During the fire, a check must be made to ensure that an adequate water supply is available to support the sprinkler system.

sprinkler systems recommend a pumping pressure of 150 psi, but department prefire planning and guidelines may modify this figure. After pumping begins, the operator must monitor the discharge pressure carefully. The pressure will tend to decrease as more heads open and may require readjustment from time to time.

If necessary, additional supply lines should be laid to the sprinkler siamese connection. Fire fighters should be prepared for an offensive attack, and hose lines should be in place to support this effort. These hose lines may be needed to conduct a search and rescue operation, protect exposures, or be used if the building's sprinkler system is not working or is inadequate. In addition, hose lines may be needed during overhaul after the sprinkler system has been shut down.

During the fire, a check must be made to ensure that an adequate water supply is available to support the sprinkler system. An additional source of water may need to be acquired to sustain the operation. If hose lines are being used in the operation, they should be supplied from a different water system than the one supplying the sprinklers.

Exposure Protection

Sprinkler systems are as effective for exposure protection as they are for fire control, provided that they are supplied with enough water. For this reason, pumpers should be hooked into the sprinkler systems of exposed buildings as well as into the system in the fire building. Pumping into an exposure should begin when its sprinkler system is activated. This especially is important when several sprinkler systems, or sprinkler systems and pumpers, are operating from the same water source. The one deterrent to effective sprinkler system operation, other than human error, is inadequate water supply. Pumpers should be used to maintain the supply to sprinklers, not to rob them of their water.

Shutting Down the System

Shutting down a system prematurely is a common mistake made when automatic sprinklers are in operation. This usually is done to decrease water damage inside the structure or to get a "better look" of the area involved; however, the fire may not have been extinguished. The fire fighter assigned to the valve should stay at that position and be ready to reopen it if the need arises. Depending on the occupancy, location of the fire, and the products or stock warehoused in the building, and the method in which these items are stored, the fire may not have been completely extinguished. Fire fighters must be ready to advance hose lines to extinguish any remaining fire.

Postfire Operations

After the fire is extinguished, the sprinkler system should be shut off; however, a fire fighter should remain at the control valves to quickly return the system to service if necessary.

The fire building should be overhauled in the usual manner. Fire fighters should check the area around and above the fire for any remaining fire and ensure final extinguishment. Water should be removed from areas in which it has collected in large quantities. Property conservation methods should be started to ensure the least amount of damage to the building and its valuable property. If at all possible, property conservation should be started during extinguishment. The incident commander should ensure that enough personnel are available to begin this task.

> **Key Points**
>
> Sprinkler systems are as effective for exposure protection as they are for fire control, provided that they are supplied with enough water.

> **Key Points**
>
> Shutting down a system prematurely is a common mistake made when automatic sprinklers are in operation.

> **Key Points**
>
> After the fire is extinguished, the sprinkler system should be shut off. A fire fighter, however, should remain at the control valves to quickly return the system to service if necessary.

> **Key Points**
>
> If possible, the sprinkler system should be restored to service before fire fighters leave the building.

If many heads have opened during a fire, there might not be enough replacements available, and/or the system may be too complex for the fire department to restore. The incident commander should contact the building's owner or management company. These individuals will be able to contact the proper group responsible for the maintenance and restoration of the sprinkler system. This is a standard procedure for many fire departments because of possible legal repercussion. The building's owner should take the proper steps to ensure the system is operational as soon as possible.

Occasionally, it will not be possible to place the sprinkler system in service soon after a fire. Such premises should be guarded either by building security, or the fire department should establish a fire detail.

Many sprinkler systems are equipped with alarm systems that notify a 24-hour guard, a private supervisory system, or the fire department alarm headquarters when a sprinkler head is activated. The alarm system, too, should be returned to service before fire fighters leave the scene.

Other Protection Systems

Other kinds of fire protection systems, usually operating in the same way as sprinkler systems, deliver fog, spray, foam, or some other special extinguishing agent. The systems and extinguishing agents vary with the type of building and/or its use; fire fighters should be aware of any such special systems in their area of responsibility and be thoroughly familiar with their operation.

Some of these systems require a manual operation for starting. A pump might have to be activated, a valve might have to be opened, or a water supply might have to be hooked up. Such operations normally are assigned to building personnel, but fire fighters must check to see that the operations have been performed properly and that fire protection systems are operating. As with sprinkler systems, fire department personnel must be sure to leave any fire protection system in operating condition or at least in the charge of someone responsible for its proper operation.

Probably the newest type of automatic fire protection system is the relatively low-cost, lightweight, quick-response sprinkler system developed in the early 1980s primarily for residential use. Residential sprinkler systems are an effective means of controlling fire in the home, allowing occupants the time to escape or be rescued. Several different systems exist, and research is ongoing to develop systems that are more efficient and cost effective.

Key Points

The newest type of automatic fire protection system is the relatively low-cost, lightweight, quick-response sprinkler system, which is designed primarily for residential use.

Wrap-Up

Chief Concepts

- An incident commander should consider a fire-suppression system inside a building as the best resource available to be used during a fire incident.
  - The two most frequently encountered fire protection systems are standpipes and automatic sprinklers.
  - Both systems may be with FDCs for fire department pumpers to deliver water to the systems if the need arises.
  - When an automatic fire protection system is present, the fire department should support the system and let it do its job.
- A standpipe system is simply a piping system in which water flows to various discharges located within a building.
  - Most standpipe systems are connected to a water source with pressure supplied by a fire pump.
- A sprinkler system is a piping system distributed throughout a building with water applied to the fire through sprinkler heads.

- Properly supplied with water, an automatic sprinkler system will usually control or extinguish a fire by itself.
- In order to be able to use fire protection systems fully, fire fighters should be aware of the location, operation, and proper use of every system in their area of responsibility.
- Standard operating guidelines should be written that address the operating procedures to be used by fire fighters working in a building with a fire suppression system.

Key Term

Standpipe system: A piping arrangement that carries water vertically and sometimes horizontally through a building for firefighting operations. It provides a means of getting water to a fire without long, time-consuming hose stretches.

1. A standpipe system supplies water _____ a building for firefighting purposes.
   a. To every stairway in
   b. Horizontally through
   c. Vertically through
   d. Horizontally or vertically through

2. Early laws required standpipes in buildings more than _____ in height.
   a. Eight stories
   b. Ten stories
   c. 75 feet
   d. 100 feet

3. The National Fire Protection Association standard for the installation of standpipes is
   a. NFPA 10
   b. NFPA 13
   c. NFPA 14
   d. NFPA 1500

4. Class _____ standpipe systems are generally intended for use by the fire department, but not for use by occupants.
   a. I
   b. II
   c. III
   d. All standpipe systems are designed for use by both fire fighters and occupants.

5. A manual-wet standpipe system
   a. Is connected to a water supply that will provide limited quantities of water for firefighting purposes
   b. Is connected to a water supply that is not capable of supplying sufficient water for firefighting purposes
   c. Is dry except for residual water in the system from testing and/or use; this system is inappropriate for areas subjected to freezing weather
   d. Standpipe systems are classified only as wet or dry; there is no manual-wet standpipe system.

6. A class III automatic wet standpipe systems must _____.
   a. Have a pressure of 65 psi and a flow capacity of 100 gpm from the most hydraulically remote hose outlet
   b. Have a pressure of 100 psi and a flow capacity of 100 gpm at the top-most hose connection of the most hydraulically remote standpipe outlet
   c. Have a pressure of 65 psi and the ability to flow two 250-gpm hose lines from the top-most and hydraulically remote hose outlet

   d. Have a pressure of 100 psi and the ability to flow two 250-gpm hose lines from the top-most and most hydraulically remote hose outlet

7. The shut-off valve for the standpipe system for a large industrial property would most likely be
   a. An outside stem-and-yoke valve
   b. An inside stem-and-yoke valve
   c. A nonindicating post valve
   d. A postindicating valve

8. If the FDC is damaged so severely that it cannot be supplied, the company assigned to supply the standpipe should
   a. Supply the standpipe through the first floor discharge outlet
   b. Supply the standpipe through the test outlets
   c. Pump into the nearest hydrant to increase the pressure
   d. Be sure the fire pump is running and lay lines up the interior stairway

9. Fire department personnel should _____.
   a. Use standpipe house lines intended for use by building occupants when they are known to be available
   b. Use lined hose provided for use by building occupants, but never used unlined hose
   c. Only use standpipe house lines intended for use by building occupants when the hose has been inspected by the fire department
   d. Never use standpipe house lines intended for use by building occupants

10. The equipment pool should be located _____ or more floors below the fire floor for a fire on an upper floor in a high-rise building.
    a. 1
    b. 2
    c. 3
    d. 4

11. When using an elevator in a high-rise building _____.
    a. Take the elevator directly to the fire floor, but not above
    b. Dismount at least one floor below the fire floor
    c. Dismount at least two floors below the fire floor
    d. Never use an elevator in a high-rise building until the fire is verified to be under control

12. Elevators equipped with fire service control
    a. Will have fire service controls in the lobby
    b. Will have fire service controls in each elevator car

c. Will have fire service controls in the lobby and in each elevator car

d. Codes vary; therefore, fire service controls locations will vary.

13. When connecting the initial attack line to a standpipe outlet in a building with standpipes in the corridors and the exact fire location is unknown, the first attack line should be connected to the standpipe outlet

   a. On the fire floor, conditions permitting
   b. One floor below the fire floor
   c. One floor or landing above the fire floor
   d. Two or more floors below the fire floor

14. A properly operating sprinkler system

   a. Will effectively control most fires, but is less effective than an attack hose being operated by a well-trained crew of fire fighters
   b. Will effectively control most fires and is more effective than a manual attack conduced by the fire department
   c. Will control the fire approximately 50% of the time
   d. Is nearly 100% effective; thus, the fire department need only reset the system and overhaul after the system achieves fire control

15. The most common type of sprinkler system is the _____ system.

   a. Wet-pipe
   b. Dry-pipe

c. Preaction
d. Deluge

16. The _____ sprinkler system has air in the sprinkler piping until a valve is opened.

   a. Dry-pipe
   b. Preaction
   c. Deluge
   d. All of the above are filled with air.

17. _____ sprinkler systems must be provided with an FDC.

   a. Wet-pipe
   b. Dry-pipe
   c. Deluge
   d. All of the above

18. When a sprinkler system is equipped with an FDC, a pumper should

   a. Hook up to the FDC and immediately pump into the system at 150 psi
   b. Hook up to the FDC and immediately pump into the system at 100 psi
   c. Hook up to the FDC and immediately pump into the system at a pressure predetermined in the building preplan
   d. Hook up to the FDC and prepare to pump into the system

Overhaul

Chapter 11

Learning Objectives

- Recognize the purpose of overhaul after the main body of fire has been extinguished.

- Assess the importance of a preinspection of a building to determine whether it is safe for fire fighters to perform overhaul tasks and duties.

- Determine the selection process and control of fire fighters during overhaul to ensure their safety and well-being.

- Identify the basic principles of overhaul to ensure that the fire is completely extinguished.

Overhaul is the task of systematically examining the aftermath of a fire scene to determine whether there is any fire, sparks, or embers remaining that could cause the reignition or rekindle of material after the fire department has left the scene. Overhaul should be accomplished in a timely manner, but there is no need to take unnecessary chances after the fire has been brought under control.

The incident commander (IC) must develop an organized and safe plan for conducting overhaul operations. Competent personnel should preinspect the building before fire fighters begin the task of overhaul. Structural integrity should determine whether firefighting activities should be conducted in all areas or designated areas of the building or whether the building should be placed off limits. Areas that are hazardous and off limits must be identified. All personnel should be aware of where they can and cannot go. Barrier tape could be used to identify and block access to these areas.

Fire fighters cannot be safe unless the scene is safe. Fresh crew should be used, relieving fire fighters who had worked during the extinguishment operations. If overhaul operations are extended, then additional or rehabbed fire fighters should be assigned.

All members should be in full PPE, including SCBA, while performing these duties. They should not be allowed to dress down. The atmosphere in the areas where overhaul operations are being performed should be monitored to determine carbon monoxide levels as well as any other toxic materials, including asbestos. This will ensure that fire fighters are wearing face pieces and are breathing from the system if levels are above the recommended levels. Personnel should be assigned in sufficient numbers and have the needed tools and equipment, including the proper number of hose lines to perform their job. Fire fighters should be well supervised, and safety officers should be assigned depending on the size and scope of the operation.

Overview

After the main body of a fire is extinguished, the fire building still might contain sparks, embers, or small-concealed areas of fires. The overhaul operation essentially is a careful and systematic examination of the fire scene. **Figure 11-1** shows fire fighters inspecting the fire building after the fire has been extinguished for any signs of remaining fire. The main purpose of overhaul is to make certain that no trace of fire remains to rekindle after the fire force has left. A second purpose is to leave the structure in as safe a condition as possible. Other individuals such as fire investigators, utility personnel, and property owners may need access to the building. Depending on the circumstances, the building could be partially occupied soon after the fire.

Cleaning up, although good for public relations, is not a necessary part of overhaul. Only debris that present a dangerous condition, such as burned structural material, furnishings, or equipment, should be removed from the building. Building components or debris that fire investigators need to inspect should be left for examination if possible. If conditions warrant and if it is safe to do so, a building's maintenance force could perform cleanup with the IC's permission.

During fire attack and related property-conservation operations, fire fighters work quickly and may need to take calculated risks occasionally. On the other hand, the fire building is overhauled after the emergency is over, and there is absolutely no reason for rushing or for taking unnecessary chances. Overhaul should be completed systematically, and the IC must develop an organized and safe plan for overhauling the fire building.

In spite of the unhurried atmosphere that should prevail during overhaul, the injury rate for fire fighters during this operation is relatively high. This chapter begins with a discussion of procedures that could help reduce the frequency of injury during overhaul: preinspection of the area to be overhauled, assignment of additional or fresher personnel to overhaul duties, along with accountability and control of fire fighters performing these tasks. This chapter then discusses the overhaul operation itself.

Preinspection

Before any firefighting personnel are sent into the building for overhaul operations, competent personnel that the IC assigned must check the fire area thoroughly. The building has been damaged and may have suffered structural damage from the fire itself or from the weight of the water used during fire attack and extinguishment. There might be holes in floors and on the roof. Stairways may be damaged and may not be safe to use. Portions of the building may be unsafe to enter under any condition. Fire fighters must be made aware of these conditions before the task of overhauling is begun.

Key Points

The overhaul operation essentially is a careful and systematic examination of the fire scene to make certain that no trace of fire remains to rekindle after the fire force has left.

Key Points

Before any firefighting personnel are sent into the building for overhaul operations, competent personnel that the IC assigned must check the fire area thoroughly.

Figure 11-1 During overhaul operations, fire fighters inspect ceilings and walls to look for any signs that a fire might not be completely extinguished.

The extent of the preinspection, as well as the entire overhaul operation, will depend on the extent of the fire. If the fire was small, confined to a room or two with contents, an inspection of the fire area, as well as the rooms above and below, should be sufficient. If a large area of the building was involved, then a more thorough inspection of the structure will need to be conducted. In many cases, the entire building will need to be examined.

The reason for inspecting is simple: The building must be safe for fire fighters to perform their tasks. Safety officers should ensure that unsafe areas are marked or taped off and that structurally unsafe areas of the building are placed off limits. Barrier tape is useful in identifying and blocking off areas that are unsafe. Fire fighters working in these areas must be well supervised. Portable lighting should be supplied to areas that

are in the dark. **Figure 11-2** shows an unsafe area that has been illuminated with additional lighting and clearly marked to warn fire fighters of damaged stairways and holes in the floor.

Fire fighters should be in full protective clothing and breathing from SCBA face pieces while overhauling **Figure 11-3**. They should not be allowed to dress down during overhaul. Carbon dioxide levels have been shown to be higher during overhaul than during actual firefighting activities. In addition, there may be asbestos and/or other damaging materials that could harm unprotected fire fighters. Debris with sharp and jagged edges, glass, hot surfaces, and other hazards abound during overhaul. Fire fighters must be in full PPE, be aware of their surroundings, and complete their task in a systematic manner.

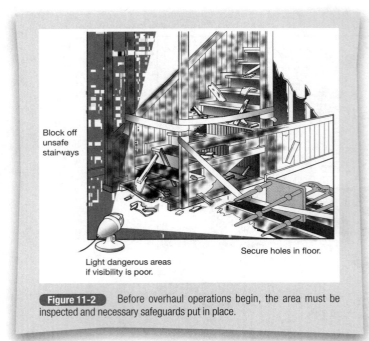

Block off unsafe stairways

Light dangerous areas if visibility is poor.

Secure holes in floor.

Figure 11-2 Before overhaul operations begin, the area must be inspected and necessary safeguards put in place.

Figure 11-3 Fire fighters without protective clothing should not be allowed in the building.

During overhaul, fire fighters should check the area for possible clues to the ignition of the fire. Indiscriminately tossing everything out the window and washing down everything in site will probably negate any chance of fire investigators determining the cause of a suspicious fire. A fire of incendiary origin can greatly alter overhaul operations or delay them until investigators can examine the fire scene. After this is done, the basic overhaul operation can begin.

Personnel

Personnel who have been fighting a fire for some time are tired and worn down by the physical activity, heat, smoke, and the weight of their equipment. These individuals generally are not in shape to conduct a careful examination of the fire premises or immediately begin overhaul operations. They deserve a rest before reassignment, and the beginning of an extensive overhaul operation is a good time to have one while the building is being inspected. These fire fighters should be sent to a rehabilitation sector to rest, treat injuries, replace fluids, and have basic vital signs checked **Figure 11-4**. It then should be determined whether they are fit to return to duty. If these fire fighters are needed later, they will be rested, alert, and better able to perform their duties.

Fire fighters who were assigned to staging manned unused backup lines or who were on exposure assignments during fire attack might be assigned to overhaul. Personnel arriving independently, such as off-duty paid fire fighters or additional arriving volunteers, are candidates for overhaul duties. If necessary, additional engine and/or ladder companies can be called to the fire scene for overhaul.

Figure 11-4 Crew who worked during the initial attack phase should be allowed time to rest.

Control of Overhaul Personnel

Crew members who have not been inside the fire building during the firefighting operations will not know which areas have been damaged or where the structure may be weak. The outside appearance of the building might give them little information about the interior. It is, therefore, extremely important that overhaul personnel be well supervised within the structure. Safety officers should be assigned depending on the size and scope of the operation.

Fire fighters performing overhaul should be formed into groups that should be assigned to a sector officer and to a certain area of the building. Members working within the sector must maintain contact with each other and with the officer in charge of the overhaul operation. No one should be allowed inside the fire building without first reporting to the sector officer assigned to the area in which they will be working. Fire fighters entering

the building without reporting could find themselves in serious trouble if no one is aware that they are inside.

The reason for these procedures is clear. The fire may have been extinguished, but there is a high probability that areas of the building could be unsafe. Following standard operating guidelines will minimize the chance of an accident, minimize the chance of serious injury should an accident occur, and ensure accountability of members working inside the building.

Work Assignments

Ordinarily, ladder companies are assigned tasks involving hand and power tools during overhaul. Engine companies advance and operate hose lines for extinguishing small fires and residual embers. If ladder company personnel are not on scene or are engaged elsewhere on the fire ground, engine company personnel will have to perform both duties. In this case, the sector officer should assign part of the overhaul group to perform ladder company tasks and the remainder of the group to perform as an engine company. Each fire fighter then knows what task he or she has been assigned.

If the group must be split up to perform its duties, there is an obvious way to make the split. Ladder company crews are usually sent to the floors above and below the fire to check for fire extension as yet undetected. Engine company crews standing by with hose lines are called to these floors if a hot spot is discovered.

Jobs that are not specifically assigned as ladder company work or engine company work would be performed by both crews. If heavy material has to be removed from an area of the structure, everyone in the group would be expected to lend a hand.

Overhaul Operation

Because less water is needed for overhaul than for firefighting operations, 2½-inch attack lines can be reduced to 1¾-inch attack lines for easier handling. If companies are released from the incident, hose lines should be cleared from the streets to a more convenient location so that returning fire apparatus will not have to run over them and traffic flow can be restored in the area.

The basic purpose of the overhaul operation is to make sure that the fire is extinguished completely. This means that any area that could possibly have been in contact with fire or intense

Figure 11-5 Fire fighters must look for signs of hidden fire, such as a wall that is hot to the touch, visible smoke, smoke patterns or discoloration on the wall, and paint blisters.

Figure 11-6 When fire is suspected behind walls or ceilings, they should be opened until a clean area is found. The area should be wet down with a light stream if necessary.

heat should be checked, whether the building is fire resistant or not. Often, although building materials resist fire, vertical or horizontal channels, shafts, and voids make perfect avenues for fire spread. During overhaul, which begins close to the area where firefighting operations ended, it is essential to look for signs of hidden fire **Figure 11-5**.

What to Look for and Where to Look

Overhaul should begin close to the area where firefighting operations ended. Fire fighters should listen for the sound of fire, beware of high-heat areas, and look for flames and smoke in the area in which they are working. If enough heat is generated, the use of thermal imaging cameras enables fire fighters to detect a hot spot on or behind a surface. During overhaul, the camera allows fire fighters to identify hazards that may not otherwise be seen. They can open spaces where fire has been detected while leaving uninvolved areas intact. Concealed horizontal and vertical spaces, walls, and ceilings should be opened and inspected if it is suspected that there is fire within. Areas that were opened during fire attack should also be rechecked. Portable lighting will be useful when checking concealed spaces. If walls and ceilings have been in contact with fire and heat, they must

be opened and checked for signs of fire. Any suspicious area should be wet down **Figure 11-6**.

Ceiling spaces must be thoroughly examined. Any fire in this area will be guided to wall spaces and then through the building vertically. False ceiling also must be checked and opened if necessary. Ceilings should be checked thoroughly while attempting to minimize damage. Property conservation should be practiced during overhaul operations. If possible, remove or collect furnishings or stock. Salvage covers can be used to cover and protect items that are unable to be removed. Every effort should be made to safeguard items in the building from further damage.

A check also should be made to determine whether sparks or embers have been carried into walls or partitions. If a sign of smoke or a strong odor exists, a light stream should be directed into the area to wet it. Fire fighters should look for steam if a hot spot has been encountered. An area in which cobwebs are found probably has not seen the effects of heat and smoke; cobwebs tend to shrivel up in temperatures that are higher than normal.

If a wall, ceiling, floor, or any other shaft shows signs of fire damage on examination, that area should be opened further until the full extent of the fire damage is visible. A hose line should be

Key Points

Vertical or horizontal channels, shafts, and voids make perfect avenues for fire spread. These areas must be thoroughly checked.

Key Points

Ceiling spaces must be thoroughly examined for hidden fire. A check should also be made to determine whether sparks or embers have been carried into walls or partitions.

available to extinguish the fire as the full area of the concealed area is exposed. All areas checked during overhaul must be checked again before this task is considered completed.

If the building has been insulated and fire has involved areas around the insulation, this material must be checked. The insulation must be removed in all areas of fire involvement until there is no indication of fire damage. Blown-in insulation is difficult to work with, but fire fighters must ensure that there has been no fire extension.

Checking Above the Fire

Checking above the fire is carried out in a similar fashion as on the fire floor. In addition, baseboards may have to be removed for a positive check of fire travel through walls and partitions. Any doubtful area should be opened up further to expose fire extension and wet down.

When fire has penetrated a ceiling space on the floor below, it might have also extended into the floor above. A full examination of this area may require the removal of flooring. This especially is important along walls and partitions and where shafts or other vertical spaces pass through the floor. If part of the floor must be removed, it should be taken up until a clean area shows the full extent of the fire. In order to hold damage to a minimum, the flooring should be removed as cleanly as possible.

Depending on department standard operating guidelines and availability of equipment, cutting floors with power saws usually is faster and cleaner than using hand tools. In general, power saws are preferred over axes and other hand tools during cutting operations.

Vertical Shafts and Channels If it is apparent or suspected that fire has spread into a vertical shaft or channel, the space must be opened and checked. If openings into the space are available, they should be used for this purpose; otherwise, the space should be physically opened. The space should be examined for signs of fire and smoke, and when necessary, a stream or streams should be directed up or down into the area to wet down possible hot spots. Fire fighters ordinarily should be assigned to check the bottom and the top of the shaft for fire and sparks.

Shafts that were opened or vented during fire control operations must be checked out thoroughly at this time even though the fire apparently has been extinguished because the

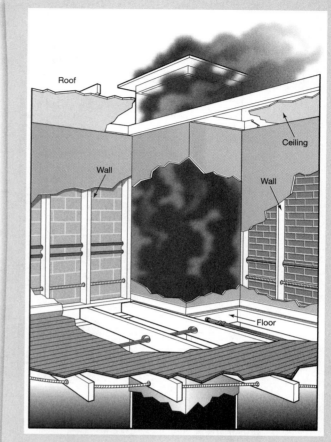

Figure 11-7   When a shaft has been the avenue of travel for a fire, the areas adjacent to the shaft must be opened.

intense heat that was confined within the shaft might have rekindled the fire. For the same reason, anything in contact with these shafts, such as floors, walls, attic spaces, or the roof, must be checked thoroughly **Figure 11-7** .

Cabinets and Compartments   When fire has traveled definitely in walls and enclosed spaces or penetrated ceiling and floor spaces, cabinets, including base and wall cabinets, in contact with these areas also must be thoroughly checked. This especially is important when cabinets, such as those in residential kitchens, are involved. These units are usually constructed with a 3- to

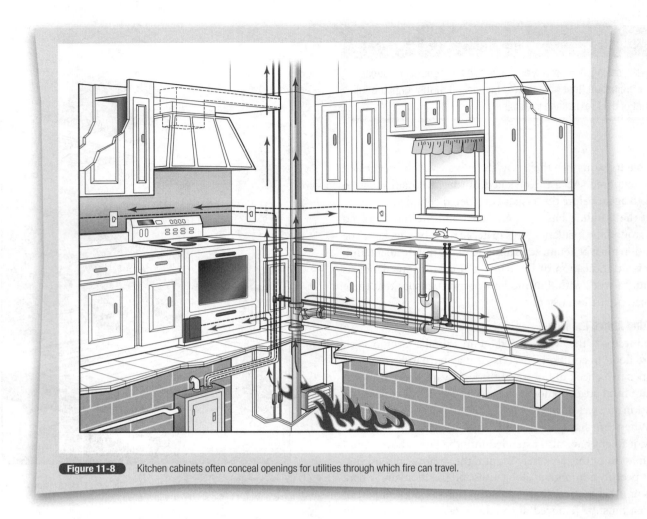

Figure 11-8 Kitchen cabinets often conceal openings for utilities through which fire can travel.

5-inch enclosed space between the floor and the bottom shelf. In many cases, this enclosed space carries electrical conduits, gas pipes, water pipes, or drains and may be open to a wall or a shaft; fire can be overlooked in such a concealed space if it is not checked **Figure 11-8** . Areas behind appliances, such as stoves, refrigerators, washers and dryers, also should be checked for fire extension.

Windows and Door Casings   When fire has involved window and door casings, they must be removed and the concealed recesses checked for fire **Figure 11-9** . The wall will need to be opened and wet down if it appears that fire has traveled into this area. The area should be opened until there is no sign of fire extension. Wainscoting is handled similarly; it is removed until a clean area is found and the area is wet down as necessary.

Water Removal

During overhaul operations, excess water must be removed from the building. In some instances, this must be completed first to reduce the weight on the floors so that they are safe to walk on. Water can be pushed and swept down vertical openings such as shafts, stairways, soil pipes opened at floor level, and existing drains. When the building is heavily damaged and the water is deep, a hole can be made in outside walls, from window sills to floor level, through which water can drain to the outside.

Key Points

During overhaul operations, excess water must be removed from the building.

Portable pumps also can be used to remove water from low levels. If enough water has accumulated in a basement or other area to endanger fire fighters if they enter, a safety officer should be assigned to that sector. The area should be marked, and all fire fighters should be made aware of the danger.

In removing water from a structure, as in all facets of overhaul, care should be taken to cause as little damage as possible. There is little point in doing a good firefighting job and then unnecessarily damaging the building after the fire is out.

Key Points

In removing water from a structure, as in all facets of overhaul, care should be taken to cause as little damage as possible.

Figure 11-9 Window and door casings must be removed to check for hidden fire.

Wrap-Up

Chief Concepts

- Overhaul is the task of systematically examining the aftermath of a fire scene to determine whether there is any fire, sparks, or embers remaining that could cause the reignition or rekindle of material after the fire department has left the scene.
  - Overhaul, if performed correctly, will assist in the fire investigation, protect the building from the elements, and return the building to the owner in as safe a condition as possible.
- Overhaul operations should begin soon after the fire has been brought under control.
  - The operation should be conducted in a timely manner without the need of taking unnecessary chances to fire fighters performing the task.
  - There should be a preinspection of the building by competent personnel before fire fighters begin overhaul to determine whether it is safe to proceed.
  - All personnel must know where they can and cannot go.
  - Areas that are off limits must be identified, and access to these areas must be blocked.
  - Fresh crews should relieve the fire fighters who were engaged in extinguishment.
  - Following standard operating guidelines will minimize the chance of an accident or serious injury.

- Fire fighters should be well supervised, and safety officers should be assigned depending on the size and scope of the operation.
- Fire fighters must open and check any area that possibly could have had contact with fire or intense heat.
  - Walls, ceilings, and floors may need to be opened.
  - These areas conceal horizontal and vertical channels, shafts, and voids and make perfect avenues for fire spread.
- If a wall ceiling, floor, or any other area shows signs of fire damage on examination, that area should be opened further until the full extent of the fire damage is visible.
- Fire fighters should be mindful of any water damage or the accumulation of water within the building.
- During overhaul operations, excess water must be removed from the building.

Key Term

Overhaul: Examination of all areas of the building and contents involved in a fire to ensure that the fire is completely extinguished.

1. During the _____ phase of a fire operation, fire fighters need not take calculated risks.
   a. Life safety
   b. Extinguishment
   c. Overhaul
   d. All of the above

2. When conducting a preinspection before overhaul
   a. Always check the entire structure for signs of fire extension.
   b. Always check the attic or cockloft area.
   c. Always check the immediate fire area, floor above, and attic/cockloft.
   d. The extent of the inspection will depend on the extent of fire.

3. Whenever possible, overhaul should be performed by
   a. The first-arriving companies who are most familiar with the interior because of their involvement in the interior attack and rescue operation
   b. Later-arriving units assigned to search and rescue
   c. Personnel who were not involved in the interior operation
   d. Any unit that has been adequately rested

4. During the overhaul phase of the operation
   a. It is extremely important that personnel involved in overhaul be well supervised.
   b. Company level supervision should be adequate.
   c. The level of supervision can be reduced because of the nonemergency nature of the operation.
   d. The incident management system, RIT, and accountability can be discontinued.

5. During overhaul
   a. Engine companies advance and operate hose lines.
   b. Ladder companies use tools.
   c. Engine companies can be assigned to operate tools and/ or hose lines.
   d. All of the above

6. When opening and checking inside a concealed space, the presence of cobwebs
   a. Is a poor indicator of the possible presence of fire, although many in the fire service still believe them to be an indicator that fire has not extended to that area
   b. Is considered one of many signs that fire has not entered the space, but should not be relied on
   c. Indicates that the area probably has not seen the effects of heat and smoke
   d. Is positive proof that fire has not entered the concealed space

7. If part of the floor must be removed during overhaul operations, it is best to use a(n) _____ to remove the flooring.
   a. Power saw
   b. Ax
   c. Hand saw
   d. Halligan tool

8. Excess water should be removed during overhaul. To remove the water
   a. Sweep it to the stairway.
   b. Use a squeegee to push it to a soil pipe opening at floor level.
   c. Place a hole in the outside wall from window sill to floor level.
   d. All of the above

Appendix

Imperial and Metric Conversions

Table A-1 Length

1 inch = 0.08333 foot, 1,000 mils, 25.40 millimeters

1 foot = 0.3333 yard, 12 inches, 0.3048 meter, 304.8 millimeters

1 yard = 3 feet, 36 inches, 0.9144 meter

1 rod = 16.5 feet, 5.5 yards, 5.029 meters

1 mile = (U.S. and British) = 5,280 feet, 1.609 kilometers, 0.8684 nautical mile

1 millimeter = 0.03937 inch, 39.37 mils, 0.001 meter, 0.1 centimeter, 100 microns

1 meter = 1.094 yards, 3.281 feet, 39.37 inches, 1,000 millimeters

1 kilometer = 0.6214 mile, 1.094 yards, 3,281 feet, 1,000 meters

1 nautical mile = 1.152 miles (statute), 1.853 kilometers

1 micron = 0.03937 mil, 0.00003937 inch

1 mil = 0.001 inch, 0.0254 millimeters, 25.40 microns

1 degree = 1/360 circumference of a circle, 60 minutes, 3,600 seconds

1 minute = 1/60 degree, 60 seconds

1 second = 1/60 minute, 1/3600 degree

Table A-2 Area

1 square inch = 0.006944 square foot, 1,273,000 circular mils, 645.2 square millimeters

1 square foot = 0.1111 square yard, 144 square inches, 0.09290 square meter, 92,900 square millimeters

1 square yard = 9 square feet, 1,296 square inches, 0.8361 square meter

1 acre = 43,560 square feet, 4,840 square yards, 0.001563 square mile, 4,047 square meters, 160 square rods

1 square mile = 640 acres, 102,400 square rods, 3,097,600 square yards, 2.590 square kilometers

1 square millimeter = 0.001550 square inch, 1.974 circular mils

1 square meter = 1.196 square yards, 10.76 square feet, 1,550 square inches, 1,000,000 square millimeters

1 square kilometer = 0.3861 square mile, 247.1 acres, 1,196,000 square yards, 1,000,000 square meters

1 circular mil = 0.7854 square mil, 0.0005067 square millimeter, 0.0000007854 square inch

Table A-3 Volume (Capacity)

1 fluid ounce = 1.805 cubic inches, 29.57 milliliters, 0.03125 quarts (U.S.) liquid measure

1 cubic inch = 0.5541 fluid ounce, 16.39 milliliters

1 cubic foot = 7.481 gallons (U.S.), 6.229 gallons (British), 1,728 cubic inches, 0.02832 cubic meter, 28.32 liters

1 cubic yard = 27 cubic feet, 46,656 cubic inches, 0.7646 cubic meter, 746.6 liters, 202.2 gallons (U.S.), 168.4 gallons (British)

1 gill = 0.03125 gallon, 0.125 quart, 4 ounces, 7.219 cubic inches, 118.3 milliliters

1 pint = 0.01671 cubic foot, 28.88 cubic inches, 0.125 gallon, 4 gills, 16 fluid ounces, 473.2 milliliters

1 quart = 2 pints, 32 fluid ounces, 0.9464 liter, 946.4 milliliters, 8 gills, 57.75 cubic inches

1 U.S. gallon = 4 quarts, 128 fluid ounces, 231.0 cubic inches, 0.1337 cubic foot, 3.785 liters (cubic decimeters), 3,785 milliliters, 0.8327 Imperial gallon

1 Imperial (British and Canadian) gallon = 1.201 U.S. gallons, 0.1605 cubic foot, 277.3 cubic inches, 4.546 liters (cubic decimeters), 4,546 milliliters

1 U.S. bushel = 2,150 cubic inches, 0.9694 British bushel, 35.24 liters

1 barrel (U.S. liquid) = 31.5 gallons (various industries have special definitions of a barrel)

1 barrel (petroleum) = 42.0 gallons

1 millimeter = 0.03381 fluid ounce, 0.06102 cubic inch, 0.001 liter

1 liter (cubic decimeter) = 0.2642 gallon, 0.03532 cubic foot, 1.057 quarts, 33.81 fluid ounces, 61.03 cubic inches, 1,000 milliliters

1 cubic meter (kiloliter) = 1.308 cubic yards, 35.32 cubic feet, 264.2 gallons, 1,000 liters

1 cord = 128 cubic feet, 8 feet × 4 feet × 4 feet, 3.625 cubic meters

Table A-4 Weight

1 grain = 0.0001428 pound

1 ounce (avoirdupois) = 0.06250 pound (avoirdupois), 28.35 grams, 437.5 grains

1 pound (avoirdupois) = the mass of 27.69 cubic inches of water weighed in air at 4°C (39.2°F) and 760 millimeters of mercury (atmospheric pressure), 16 ounces (avoirdupois), 0.4536 kilogram, 453.6 grams, 7,000 grains

1 long ton (U.S. and British) = 1.120 short tons, 2,240 pounds, 1.016 metric tons, 1016 kilograms

1 short ton (U.S. and British) = 0.8929 long ton, 2,000 pounds, 0.9072 metric ton, 907.2 kilograms

1 milligram = 0.001 gram, 0.000002205 pound (avoirdupois)

1 gram = 0.002205 pound (avoirdupois), 0.03527 ounce, 0.001 kilogram, 15.43 grains

1 kilogram = the mass of 1 liter of water in air at 4°C and 760 millimeters of mercury (atmospheric pressure), 2.205 pounds (avoirdupois), 35.27 ounces (avoirdupois), 1,000 grams

1 metric ton = 0.9842 long ton, 1.1023 short tons, 2,205 pounds, 1,000 kilograms

Table A-5 Density

1 gram per millimeter = 0.03613 pound per cubic inch, 8,345 pounds per gallon, 62.43 pounds per cubic foot, 998.9 ounces per cubic foot

Mercury at 0°C = 0.1360 grams per millimeter basic value used in expressing pressures in terms of columns of mercury

1 pound per cubic foot = 16.02 kilograms per cubic meter

1 pound per gallon = 0.1198 gram per millimeter

Table A-6 Flow

1 cubic foot per minute = 0.1247 gallon per second, 0.4720 liter per second, 472.0 milliliters per second = 0.028 m^3/min, lcfm/ft^2 = 0.305 m^3/min/m^2

1 gallon per minute = 0.06308 liter per second, 1,440 gallons per day, 0.002228 cubic foot per second

1 gallon per minute per square foot = 40.746 mm/min, 40.746 l/min · m^2

1 liter per second = 2.119 cubic feet per minute, 15.85 gallons (U.S.) per minute

1 liter per minute = 0.0005885 cubic foot per second, 0.004403 gallon per second

Table A-7 Pressure

1 atmosphere = pressure exerted by 760 millimeters of mercury of standard density at 0°C, 14.70 pounds per square inch, 29.92 inches of mercury at 32°F, 33.90 feet of water at 39.2°F, 101.3 kilopascal

1 millimeter of mercury (at 0°C) = 0.001316 atmosphere, 0.01934 pound per square inch, 0.04460 foot of water (4°C or 39.2°F), 0.0193 pound per square inch, 0.1333 kilopascal

1 inch of water (at 39.2°F) = 0.00246 atmosphere, 0.0361 pound per square inch, 0.0736 inch of mercury (at 32°F), 0.2491 kilopascal

1 foot of water (at 39.2°F) = 0.02950 atmosphere, 0.4335 pound per square inch, 0.8827 inch of mercury (at 32°F), 22.42 millimeters of mercury, 2.989 kilopascal

1 inch of mercury (at 32°F) = 0.03342 atmosphere, 0.4912 pound per square inch, 1.133 feet of water, 13.60 inches of water (at 39.2°F), 3.386 kilopascal

1 millibar (1/1000 bar) = 0.02953 inch of mercury. A bar is the pressure exerted by a force of one million dynes on a square centimeter of surface

1 pound per square inch = 0.06805 atmosphere, 2.036 inches of mercury, 2.307 feet of water, 51.72 millimeters of mercury, 27.67 inches of water (at 39.2°F), 144 pounds per square foot, 2,304 ounces per square foot, 6.895 kilopascal

1 pound per square foot = 0.00047 atmosphere, 0.00694 pound per square inch, 0.0160 foot of water, 0.391 millimeter of mercury, 0.04788 kilopascal

Absolute pressure = the sum of the gage pressure and the barometric pressure

1 ton (short) per square foot = 0.9451 atmosphere, 13.89 pounds per square inch, 9,765 kilograms per square meter

Table A-8 Temperature

Temperature Celsius = 5/9 (temperature Fahrenheit − 32°)

Temperature Fahrenheit = 9/5 × temperature Celsius + 32°

Rankine (Fahrenheit absolute) = temperature Fahrenheit + 459.67°

Kelvin (Celsius absolute) = temperature Celsius + 273.15°

Freezing point of water: Celsius = 0°; Fahrenheit = 32°

Boiling point of water: Celsius = 100°; Fahrenheit = 212°

Absolute zero: Celsius = −273.15°; Fahrenheit = −459.67°

Table A-9 Sprinkler Discharge

1 gallon per minute per square foot (gpm/ft^2) = 40.75 liters per minute per square meter (Lpm/m^2) = 40.75 millimeters per minute (mm/min)

Glossary

Backdraft: When oxygen enters a structure that is filled with the products of combustion and contains heat and fuel, the accumulated gases may ignite into a rapidly spreading fire or a violent explosion.

Backup line: An additional hose line used to reinforce and protect personnel in the event the initial attack proves inadequate.

Ball valve: Valves used on nozzles, gated wyes, and engine discharge gates. Made up of a ball with a hole in the middle of the ball.

Combination attack: A type of attack employing both the direct and indirect attack methods.

Conduction: The travel of heat through a solid body.

Convection: The travel of heat through the motion of heated matter.

Convection cycle: Heat transfer by circulation with a medium such as a gas or a liquid.

Crosslays: Traverse hose beds.

Direct attack: Firefighting operations involving the application of extinguishing agents directly onto the burning fuel.

Divided hose bed: A hose bed that is separated into two supply hose compartments running the length of the hose bed.

Double male and double female fittings: Used to connect two threaded connections of the same size and sex.

Fire tetrahedron: A geometric shape used to depict the four components required for a fire to occur: fuel, oxygen, heat, and chemical chain reactions.

Flashover: Ignition of combustibles in an area heated by convection, radiation, or a combination of the two.

Hydrant assist valve (HAV): Also known as a four-way valve.

Incident Management System (IMS): An organized system of roles, responsibilities, and standard operating guidelines used to manage emergency operations.

Indirect attack: Firefighting operations involving the application of extinguishing agents to reduce the buildup of heat released from a fire without applying the agent directly onto the burning fuel.

Master stream appliance: A large-capacity nozzle that can be supplied by two or more hose lines or fixed piping. It can flow in excess of 300 gallons per minute. Includes deck guns and portable ground monitors.

Overhaul: Examination of all areas of the building and contents involved in a fire to ensure that the fire is completely extinguished.

Preincident plan: A written document resulting from the gathering of general and detailed information to be used by public emergency response agencies and private industry for determining the response to reasonable anticipated emergency incidents at a specific facility.

Preincident planning: The process used to gather information to develop a preincident plan.

Prepiped master stream appliance: A master stream appliance that has a separate discharge pipe of adequate diameter that runs from the fire pump to the appliance.

Pumper: Fire apparatus with a permanently mounted fire pump of at least 750-gpm (3,000 L/min) capacity, water tank, and hose body whose primary purpose is to combat structural and associated fires.

Pumper fire apparatus: Fire apparatus with a permanently mounted fire pump of at least 750-gpm (3,000 L/min) capacity, water tank, and hose body whose primary purpose is to combat structural and associated fires.

Radiation: The travel of heat through space; no material substance is required.

Rapid intervention team (RIT): A minimum of two fully equipped personnel on site, in a ready state, for immediate rescue of injured or trapped fire fighters.

Residual pressure: The pressure remaining in a water distribution system while the water is flowing. The residual pressure indicates how much more water is potentially available.

Sizeup: Basis on which engine company operations are carried out.

Standpipe system: A piping arrangement that carries water vertically and sometimes horizontally through a building for firefighting operations. It provides a means of getting water to a fire without long, time-consuming hose stretches.

Static pressure: The pressure in the water pipe when no water is flowing.

Supply hose: The hose used to deliver water from a source to a fire pump.

Index

Photo Credits